3/08 (3)

WHAT IF THE MOON DIDN'T EXIST?

WHAT IF THE MOON DIDN'T EXIST?

Voyages to Earths That Might Have Been

NEIL F. COMINS

HarperCollinsPublishers

HarperCollins books may be purchased for educational, business, or sales promotional use. For information, please write: Special Markets Department, HarperCollins Publishers, Inc., 10 East 53rd Street, New York, NY 10022.

FIRST EDITION

Designed by George J. McKeon

Library of Congress Cataloging-in-Publication Data

Comins, Neil F., 1951–
    What if the moon didn't exist?: voyages to Earths that might have been / by Neil F. Comins.
        p. cm.
    Includes bibliographical references and index.
    ISBN 0-06-016864-1
    1. Solar system—Miscellanea. 2. Astronomy—Miscellanea. I. Title.
QB502.C77   1993
523.2—dc20   92-56198

93 94 95 96 97 ❖/HC 10 9 8 7 6 5 4 3 2 1

To my wife, Suzanne, with love and appreciation

CONTENTS

ACHNOWLEDGMENTS

WHEN MY OLDER SON, JAMES, WAS FIVE HE BEGAN INUNDATING us with questions that began with "What if": What if there were no trees? What if rocks were soft? It didn't dawn on me for many months that this type of question holds enormous potential for understanding nature. That finally happened in the fall of 1990, when my colleague David Batuski commented that people are in a rut, always looking at Earth and life from the same points of view. Intrigued, I responded, "What if the moon didn't exist? What would the earth be like then?" Although we didn't have time to pursue the question together, I found myself absorbed in creating a new version of Earth, developing the concepts that became the first chapter of this book. I am grateful to Dave for motivating me to explore roads I had not taken before and to my son for showing me how to enter them.

I am deeply indebted to my agent, Maria Carvainis, for her excellent work on my behalf. I also appreciate help from my friends Greg Clark, M.D. (for his comments on biological clocks), Craig Bohren, Ph.D. (for his comments on the passage of light through the air), Cliff Mills (for his early and continuing support of my writing), and Cathy Donovan (for her support and logistical help). The University of Maine gave me a sabbatical to write this book. Four chapters are expanded from articles that appeared in *Astronomy* magazine: chapter 1 appeared in the February 1991 issue, chapter 4 appeared in the July 1992 issue, chapter 5 appeared in the May 1992 issue, and chapter 9 appeared in the June 1991 issue.

I am especially grateful to my wife, Suzanne, for giving me the time to write by shouldering more than her fair share of caring for our children, James and Joshua, during that time. Also, her invaluable editing took many of the rough edges off the manuscript.

INTRODUCTION

EVEN AS LATE AS 1944, THE EARTH WAS ASSUMED TO BE SO MASSIVE and resilient that humans felt few qualms about dropping millions of pounds of bombs on it. Those who bothered to think about it back then reasoned that after World War II the scars created by the bombs could be healed and life would continue. Virtually everyone took the earth's resources and its ability to cope with humans for granted; human activities such as collecting raw materials, manufacturing, farming, and disposing of refuse were done with little thought to global consequences. After all, the earth seemed so big and diverse that it could surely absorb the impact of a few billion tiny humans running around on its enormous surface.

The unleashing of nuclear power in 1945 abruptly changed our relationship with the earth and its other inhabitants. We could, at the press of a "button," alter or even end the earth's capacity to support life. The use of nuclear energy demanded our attention as the first human activity with clearly global consequences. Others followed.

Throughout the 1950s and 1960s our connection to the

earth shifted even more as our ability to redirect the course of global events came into sharp focus. Indeed, an awareness of just how much humans were affecting the planet began seeping into the collective consciousness: Water and air were becoming profoundly polluted, the skies were dangerously crowded with airplanes, roads were overloaded with cars, cities were expanding rapidly and without benefit of forethought, toxic materials were being dumped and their locations forgotten, and the earth was being laid bare in strip mines and the clear-cutting of forests. Passengers traveling the oceans on steamships saw floating refuse stretching from one continent to another.

One could argue that the honeymoon period between humans and the earth finally ended in the summer of 1969, when the spacecraft *Eagle,* carrying Neil Armstrong and Buzz Aldrin, landed on the moon and the two astronauts turned their cameras on the earth. It was then that our perception of our planet telescoped graphically and permanently, and what had seemed to us limitless and unassailable now appeared small and fragile.

Our world is elegant when seen from afar. Under ever-changing white clouds, bright blue oceans are juxtaposed against brown continents. The earth's surface is vibrant and dynamic, unlike the drab, unchanging moon. The contrast between the earth and moon is particularly vivid in the photographs taken from space. When seen from across the abyss of empty space, the ravages to the earth caused by humans are invisible. When viewing pictures of the whole earth, it is easy to be lulled into thinking of our planet as an everlasting haven.

But from up close, it is clear that Earth's life-supporting capabilities are greatly stressed by human activity. We are changing the chemistries of the soil, air, and water, killing the forests by cutting down or poisoning trees, taking over the habitats of many animals and plants. Our survival and that of many other species now requires us to consider how we are affecting the earth. The monumental effort of fully understanding our impact is being carried out by people in many fields including geology, ecology, meteorology, botany, biology, chemistry, physics, and astronomy. An underlying goal of this effort, and the one on which this book is focused, is to understand what has made the earth so successful in supporting life.

We will take a novel tack in this endeavor by asking what the world would be like if its astronomical environment were different. By creating and exploring alternative earthlike worlds, we will develop a new perspective on our own planet. In each of the first five chapters we will posit an astronomically plausible change during the beginning of the solar system. For example, our first new world is identical to the earth except that it has no moon. (The changes do not propagate through the book—for example, we put the moon back in orbit before changing the mass of the sun.)

We will then create "what if" scenarios that might affect the earth in the future. We will consider what would happen to the earth in the wake of particular astronomical events, such as a nearby star exploding or entering the solar system. Next we will consider how our world would look if our eyes saw it differently. In the last chapter we will discuss the process of creating "what if" situations, then explore one case in which we humans have brought about a global change to the earth: We will examine what will happen if the ozone layer continues to be depleted.

Could the earthlike worlds and their solar systems described in the first five chapters exist? The answer is buried in the actual formation of the solar system. Our solar system (the sun, planets, moons, asteroids, and comets) formed 4.57 billion years ago. Shortly before then a star with more than four times as much mass as the sun exploded somewhere in the Milky Way galaxy. This super-nova explosion was so powerful that for several weeks it emitted nearly as much light as all the rest of the stars in the galaxy combined.

Most of the elements that exist on Earth were formed in that awesome explosion. The outer layers of the dying star, newly enriched in iron, gold, silver, nickel, cobalt, and cop-per, among many other elements, expanded into space. Eventually that gas and dust encountered similar matter floating in interstellar space. The colliding gases slowed one another, forming a giant interstellar cloud. A piece of that cloud eventually became dense enough to collapse under the influence of its own gravity. The collapsing cloudlet contained 99 percent hydrogen and helium (the simplest elements formed at the beginning of the universe) and 1 percent of the heavier elements.

The cloudlet's collapse was extremely chaotic; we now know that *small changes in the motion and mass of the super-nova remnant and in the cloud that formed from it would have led to extremely different physical conditions in the collapsing gas and dust cloudlet.* Even a microscopic change in the ini-tial motion of the interstellar matter that formed the solar system could have led to many of the changes described in this book.

In many parts of the natural world tiny changes lead to massive ones. This process is now graced with the name "butterfly phenomenon," based on the idea that the tiny wisps of wind created by the fluttering of a butterfly's wings in Rio de Janeiro can create stronger and stronger winds that may eventually lead to a hurricane off New Jersey—or they may not. This extreme sensitivity of matter to tiny changes explains why weather predicting will never be an exact science.

Analogously, by making small changes in the collaps-ing, swirling cloudlet that became our solar system, it is entirely plausible that the planets, moons, and other matter here would all be different than they are. For example, if the initial cloudlet had slightly less rotation than it did, the sun would be more massive than it is. The planets would also have different masses, chemical compositions, and, perhaps, different locations. Other changes in the collapsing cloudlet would have led to the earth having a different mass or spinning on its axis at a different angle as it orbits the sun. The earth's global properties (such as its mass, chemical composition, and rotation rate) could all have been different, since the laws of physics and chemistry would have allowed matter to be distributed differently during the formation of our solar system.

The worlds in this book do not exist. We will develop and explore them by extrapolating from conditions on the present earth and by using plausible astronomical and geological conditions or theories. We begin our journey by considering whether the earth and life on it would have evolved as they did if the moon had never been formed.

WHAT IF THE MOON DIDN'T EXIST?
SOLON

FORMATION OF THE EARTH

THE PLANETS WERE ORIGINALLY TRILLIONS OF DUST PARTICLES ORBIT-
ing in a massive disk around the newly formed sun 4.6 billion
years ago. These dust motes collided, sometimes bouncing off
one another, sometimes bonding. Clumps formed and grew to
the size of pebbles, then boulders, then mountains. These
growing masses orbited in the region between one-third and
one hundred times the earth's distance from the sun, making
the disk 9 billion miles wide. Eventually the largest clump at
our distance from the sun came to dominate over its neighbors, drawing them into itself through its gravitational force.
This clump became the earth, while large clumps elsewhere in
the disk became nuclei for the formation of the other planets.

The whisper of planetary formation swelled to a roar as
more and more clumps, called planetesimals, plunged onto
the growing surfaces of the young earth and other planets.
Most of these planetesimals ranged from a few miles to a
few hundred miles in diameter. Their impacts created tre-

mendous heat, making the early planets seas of churning, bubbling, molten rock.

Most planetesimals ended up by colliding with one another or with one of the young planets. With each impact, the amount of free debris near each planet decreased, eventually causing the collision rates of the planetesimals to decline. The impacts never stopped completely, however: They continue even today in the form of meteors. For example, each year 70 million pounds of rock and metal fall onto the earth from interplanetary space. Perhaps fifty thousand planetesimals remain in orbit around the sun as asteroids today.

The earth formed without the moon. For millions of years, perhaps longer, only the starlight and the red glow from rivers and seas of lava illuminated the night sky. However, at the same time that the earth formed, a planetesimal the size of Mars coalesced in an unusually elliptical orbit. (Ignoring the gravitational pull of the other bodies in the solar system, the path of each planet, moon, and other body around the sun is an ellipse.) The errant planetesimal's elongated path around the sun periodically took it across the earth's orbit. Although the monster planetesimal safely avoided the young earth for millions of years, eventually they met.

THE MOON'S EMERGENCE

Throughout the millions of years before the planetesimal struck, storms swirled ceaselessly on the moonless earth. Lightning lit the barren landscape, while the air reverberated with deafening thunder claps. Volcanoes emitted rivers of lava and clouds of gas and dust. Earthquakes shook the earth's surface as the magma (molten rock) in its interior clawed at the bottom of the planet's thin crust. At that time the earth was bone-dry. Any water on its surface was immediately vaporized by the heat stored in the rocks, making the planet utterly inhospitable and lifeless.

The giant planetesimal would have first appeared as a dim light in the night sky, no bigger or brighter than a faint star. Seeming to swell in size as it approached, the planetesimal would have filled the entire sky just before it smashed into the earth at twenty-five thousand miles an hour. The collision between such large bodies even when traveling at that speed would have appeared to occur in slow motion. Although the crash was over in ten minutes, the impact generated the explosive power of a billion, trillion tons of TNT. For comparison, the nuclear bombs that leveled Hiroshima and Nagasaki each contained the equivalent of twenty thousand tons of TNT.

In a matter of minutes more than 5 billion cubic miles of the earth's outer layers, its crust and mantle, were sprayed into orbit. This shattered material formed an enormous ring around the earth. The intense heat generated by the impact also ejected some of the earth's atmosphere and some of the gases attached to rocks inside the planet. Much of the gas was carbon dioxide, which was lost to the planet forever. The entire earth convulsed as it absorbed the energy from the greatest trauma it had ever endured. Everywhere its surface buckled, cracked, and dissolved as the remains of the intruder sank toward the earth's core.

The site of the impact became an enormous sea of liquid rock so hot and bright that for hours it gave off more heat and light than the equivalent area on the surface of the sun. No trace of this event remains. First the molten rock solidified into an enormous crater. Then, over the next several hundred million years, wind and rain wore down its edges. Finally, the earth's surface began moving via tectonic-plate motion (the motion of the continents over, under, and into one another), which transported all remnants of the collision site deep into the earth's mantle. There may, however, be traces of the planetesimal deep inside the earth. Parts of it may be resting on the earth's dense core.

Meanwhile, fragments of the ring around the earth created by the impact collided with one another. High-speed impacts caused ring pieces to shatter against one another and redisperse; slower impacts enabled pieces of the ring to coalesce and grow. The moon began to assemble from the ring shards, just as the earth had formed from grains of dust ages before. Eventually the gravitational pull of one massive clump became powerful enough to attract the rest of the pieces of the ring onto it, and the moon materialized.

THE MOON'S EFFECTS ON THE EARTH

High tides and romantic, moonlit nights barely hint at the profound effect the moon has had on the earth and the evolution of life upon it. The devastating impact that led to the moon's formation billions of years ago destroyed much of the earth's young crust (its surface layer), changing forever the future of the planet's continents and oceans. The evaporation of so much carbon dioxide into outer space during the collision altered the evolution of the earth's atmosphere. The blow also altered both the earth's spin and its path around the sun.

The moon affects the earth through its gravitational attraction and through the sunlight it reflects here. The most direct gravitational effect of the moon is its influence on the ocean tides. In turn, the ocean tides pull on the moon, causing it to speed up and spiral away from the earth. These tides also slow the earth's rotation. Later in this chapter and in the next one we will explore the riddle of the moon and its movement from the earth. We will also see how the earth has slowed its rate of spin, or rotation, in response.

The moon's visual effect on the earth stems from sunlight reflected here by it. (The moon gives off no light of its own.) This so-called moonlight has brightened our nights and, as we will see later in this chapter, affected the evolution of some animals. It has also served as an invaluable timekeeper that helped early humans survive the changing seasons.

The moon's direct effects on the earth have led to secondary changes, many of which have been important to the evolution of life as we know it. For example, the earth's slower rotation rate has been essential in shaping the life cycles of all plants and animals.

Just how much can the moon's role in the earth's history be ignored? *What if the moon didn't exist?* Could life exist on Earth without it? If so, how would life be different? How much of life as we know it do we owe to the moon? In this chapter we explore some of the basic changes that would occur to the earth had the moon never formed. The changes make clear how crucial our astronomical environment is to the earth's role as a habitat for life.

MOONLESS EARTH: SOLON

We will call the moonless earth Solon, emphasizing the planet's solo journey around the sun. Assuming that the rest of the solar system is the same as it is today, and that Solon has the exact size, chemical composition, and orbit around the sun as does Earth, this chapter will examine the planet's essential astronomical, geological, and life-supporting characteristics. Life on Solon, though possible, would take hundreds of millions of years longer to evolve, and that life would have to adapt to many different geological conditions.

The Formation of Solon

The planetesimal that created the moon traveled trillions of miles over millions of years before hitting Earth. It also swept by other planetesimals as well as by the planets Mars and Venus. Its orbit was altered by the gravitational force from each body it encountered. As a result of all these variations in its path, the planetesimal finally ended up striking the earth. But it need not have met this fate.

If that planetesimal had formed in an orbit different from its actual path by only a few inches, it would not have struck the earth. Over the planetesimal's lifetime the difference between the true orbit and any other path would have been amplified by the gravitational attractions it experienced passing near other bodies. This amplification effect, discovered in the 1980s, stems from a branch of mathematics called chaos. Had it begun in a slightly altered orbit, the planetesimal would easily have been twenty-five thousand miles to one side of its true path by the time it reached the earth during that last, fateful orbit. That change, absolutely minuscule in astronomical terms, would have prevented the collision.

Even such a near miss between the earth and another body is no minor event. As it passed by, the planetesimal would be whipped into a dramatically different orbit by the gravity of the nearby earth. Depending on its new course, the planetesimal might eventually strike the sun, Jupiter, or another body, or leave the solar system forever.

DIFFERENCES BETWEEN SOLON AND EARTH

The geological history of Solon would begin to diverge from that of the earth as soon as the planetesimal failed to hit the young planet. As Solon's unpunctured crust continued to cool, rainwater would eventually collect on its surface. The water would flow downward and accumulate in the lowest reaches of the planet's surface, thereby forming the first oceans. However, these oceans would occupy different areas than the first oceans on Earth. Having avoided the planetesimal's impact, young Solon would have different contours from those of our earth, with a different distribution of continents and ocean basins.

Different Tides

The oceans on Solon would still have tides despite the absence of the moon. Driven by the gravitational pull of the sun, those tides would be one-third as high as tides on Earth today. The range between high and low tide would remain constant throughout Solon's year, unlike that of the earth, where that range varies each day with the phases of the moon. This change in the height of Earth's tides is due to competition between the gravitational forces of the moon and sun, which pull on the earth in different directions during different phases of the moon. For example, during a new moon, when the moon is between the earth and the sun, the moon and the sun pull on the oceans in the same direction and the tides are especially high. During a quarter moon (when one half of the moon's face is visible), the gravitational forces act in different directions and partially cancel each other, leading to lower tides. Just as tides do on Earth, the relatively gentle tides would change Solon's rotation rate—that is, how fast it spins on its axis.

If Solon didn't rotate, high tide would be located directly between the sun and the center of the planet. But like the young earth, young Solon would rotate once every six hours, and this would drag the water in the high tide away from the point directly below the sun. The sun would try to pull the high tide back beneath it, and as it flows back toward the sun, this tidal water would scrape against the rough ocean bottoms and encounter continents in its way. The motion of the water pushing against the continents and the friction generated by that water rubbing against the ocean bottom would both act to slow down Solon's rotation. As we will discuss in more detail in chapter 2, a similar process also occurs on Earth due to the strong gravitational pull of the moon.

This tidal dragging of the water across the surface of

Solon would create a slow but inexorable lengthening of the days. As a result, over the next 4.5 billion years Solon would spin more and more slowly until the day becomes eight hours long. For the sake of comparison, note that the greater gravitational force from the moon has lengthened the earth's day from six hours to twenty-four hours during the same period of time. Today a year on Solon would have 1,095 eight-hour days. The sun would be up for between three and five hours each day in the midlatitudes, depending on the season. Solon's rapid rotation would be a major source of its differences from the earth. The most obvious contrast between the two bodies is the amount of surface wind each would experience.

Different Winds

The total wind pattern around a planet is a complex combination of vertical and horizontal motions. Winds are generated by both a planet's rotation and the heating and cooling of its air. The rotation drags the air along the planet's surface, while heating makes air rise and cooling makes air descend. The horizontal motion of equatorial and polar winds on the earth is from east to west, whereas winds in the midlatitudes travel west to east. All these winds often meander north or south and sometimes they flow in the opposite direction from their normal course. This is why weather forecasters include the current wind direction.

Winds on Solon would be very different from those on Earth because the faster a planet rotates, the more its winds move east or west and the less they wander north or south. We can see this today on Jupiter and Saturn, both of which spin more rapidly than the earth.

Jupiter and Saturn have ten-hour days. They rotate so rapidly that the friction between their surfaces and atmospheres pulls the air into narrow, streaming belts of wind. These winds whip around the planets from east to west or west to east at speeds of up to three hundred miles an hour. The direction of the wind on Jupiter and Saturn depends on latitude (distance from the planets' equators). Frequent storms disrupt the east-west flow of air.

Powerful hurricanes sometimes last for years and even for centuries on Jupiter and Saturn, causing the air to move in circular or oval paths, just as hurricanes do on Earth. (Through a telescope you can see both the bands of winds and a storm called the Great Red Spot on Jupiter.) Similar to Jupiter and Saturn, Solon would have stronger and more persistent winds than those on Earth, and they would normally flow east to west, parallel to Solon's equator, with much less north-south motion than winds on Earth. Hurricanes on Solon would be both more powerful and more frequent than those on Earth, with winds regularly topping two hundred miles per hour.

Winds are responsible for most of the waves on Earth's oceans. Wind-driven waves are caused by friction between the wind and the water. Since Solon's winds would be so much stronger than winds on Earth, it follows that the waves on Solon would be higher and more ferocious than any encountered here except during the most severe storms.

The stronger, more persistent winds on Solon would also generate more friction between the air and the exposed mountains than that which occurs on Earth. The winds and the debris they carry would grate against the naked rock, working along with water to wear down mountains on Solon more rapidly than such erosion occurs on Earth. All other things being equal, Solon's highest mountains should be lower than the highest mountains on Earth.

Different Magnetic Fields

Solon, like the earth, would have a molten (liquid) core of dense metals that sank there from the surface when the planet formed 4.6 billion years earlier. Solon's rapid rotation combined with its molten, metallic core would also cause the planet to generate a magnetic field similar to, but stronger than, the one created by the earth. Magnetic fields are created whenever charged particles, such as electrons and ions (atoms missing one or more electron), change speed or direction.

The hot cores inside both the earth and Solon would maintain many such charged particles. These particles would move to create the magnetic fields for two reasons. First, the planets' rotations would make the charged particles move in circles. Second, the charged particles would move upward or downward in the core, depending on whether that part of the core is cooling or heating. This vertical motion, called convection, is the same as that of water boiling in a pot: Blobs of water move to the pot's surface, give up their heat, and settle back down to be reheated.

The rotating, convecting, charged magma inside Solon's core would generate a stronger magnetic field than that possessed by the earth. The faster a body rotates, the stronger its magnetic field. Therefore, Solon's magnetic field would be almost three times stronger than the earth's. Like the earth's, Solon's magnetic field would extend out from one magnetic pole, sweep over the whole planet like an invisible casing, and return inside the planet at the other magnetic pole. Planetary magnetic fields are extremely important because they deflect the flow of high-energy particles from the sun, called the solar wind, preventing them from striking the planet's atmosphere. Frequent impacts from the solar wind would alter the chemistry of a planet's atmosphere, potentially threatening the evolution of life on the planet's surface.

On Earth most of the solar wind is deflected by the earth's magnetic field into two donut-shaped regions several thousand miles above the planet's surface. These are the Van Allen radiation belts, named after James A. Van Allen, whose research team first discovered them in 1958. Solon's stronger magnetic field would present an even tougher screen around that planet than do the Van Allen belts around the earth, preventing more solar-wind particles from reaching the planet's surface. Since the earth's Van Allen belts overload with particles during periods of intense solar activity, some of these particles cascade into Earth's lower atmosphere and create the aurora borealis (northern lights) and aurora australis (southern lights). The stronger magnetic field around Solon would make it more difficult for an overload to occur there. As a result, Solon would experience fewer auroras than does the earth.

Different Early Atmospheres

The air we breathe is composed of 78 percent nitrogen, nearly 21 percent oxygen, and traces of other gases such as argon, water vapor, and carbon dioxide. Breathing is so automatic for us that most of us assume that the air we breathe has been here throughout history. It hasn't, nor would it have been on Solon. There, as on Earth, hydrogen and helium would first dominate the atmosphere. These gases were part of the disk from which Solon would form, and they would be part of its primordial atmosphere as it first coalesces into existence. But Solon's gravitational forces would be too weak to hold these elements for long. Heated by sunlight, they would expand and rise, floating higher and higher above the planet's surface until they eventually drift out into space, never to return.

Solon's second atmosphere would come from its interior, rather than from space above it. This atmosphere would be dominated by carbon dioxide, nitrogen, and water vapor, which were originally attached to the surfaces of rocks inside Solon. The heat generated by planetesimal impacts and by radioactive elements in the planet would cause the rocks to release these gases, which would then percolate their way upward. Most of this second atmosphere would finally emerge through volcanoes and cracks in the planet's surface within a few hundred million years of Solon's formation. (Such outgassing of rocks due to heating from the earth's core still occurs and is the source of the great volume of gas emitted by volcanoes on Earth today.) Since the gases comprising this second atmosphere are all heavier than hydrogen and helium, Solon's gravity would be strong enough to hold them indefinitely.

We can estimate how thick Solon's second atmosphere would become by examining the atmosphere of its sister planet, Venus. Jewel of the night sky, Venus is very similar to Solon in size, chemical composition, and distance from the sun. Venus has 82 percent as much matter (mass) and an average density 95 percent as great as Solon's. The similarity in densities tells us that the chemical composition of Venus essentially matches that of both Solon and Earth. Because of these common characteristics, astronomers look to Venus as a model of the earth's former carbon dioxide atmosphere. Following their lead, we will compare the atmospheres of Venus, Earth, and Solon.

The original outgassing of carbon dioxide from Venus gave it an atmosphere 115 times denser than Earth's at present. Since Venus and the earth are so similar in other ways, it might not come as a surprise that geologists estimate that Earth's early, carbon dioxide–dominated atmosphere was 100 times denser than our present nitrogen-oxygen atmosphere. The difference between the early atmospheres of Venus and Earth comes from the fact that Venus does not have a moon. This suggests that Venus was never struck by an atmosphere-removing, Mars-sized planetesimal.

Since Solon would not be struck either, we would expect its pristine carbon dioxide atmosphere to have the same density as that of Venus. Having said that, we are now ready to consider Solon's suitability as a habitat for complex life forms.

For Solon to support life on its continents, its carbon dioxide air would have to be converted into a breathable nitrogen-oxygen atmosphere. After all, animals on Earth need the energy stored in oxygen in order to live; we could not function in a carbon dioxide–dominated atmosphere. Only if this conversion were possible, as it was on Earth, would we expect complex life to evolve on Solon's surface.

ORIGINS OF LIFE ON SOLON

The Final Atmospheric Conversion Begins

Like the atmosphere of Earth, but unlike that of Venus, Solon's carbon dioxide atmosphere would change over billions of years. The transformation would begin with the formation of Solon's oceans. Since water absorbs enormous quantities of carbon dioxide (which is why soda in cans, pumped full of carbon dioxide, fizzes for so long), the presence of the oceans would markedly lower the density of Solon's carbon dioxide atmosphere. Starting with more carbon dioxide than the earth started with, Solon's atmosphere would have been thicker when its oceans became saturated with the gas. Reducing the amount of carbon dioxide does not, by itself, make the air breathable. The remaining carbon dioxide must be converted into oxygen.

The nitrogen component of the final nitrogen-oxygen atmosphere would already be present. Part of the early carbon dioxide–dominated atmosphere, this nitrogen would grow in importance as the amount of carbon dioxide is reduced. The oxygen would come later, with the evolution of life in Solon's oceans.

The Energy to Create Life

Evolution on Earth began less than a billion years after our planet formed. Because all the essential ingredients for the first, simple aquatic life forms would also exist on Solon, life would begin evolving there early in the planet's history. As on Earth, water would serve as the medium in which the building blocks of life—chemicals and inorganic compounds containing nitrogen, potassium, iron, and other elements washed down to the oceans by rivers and ocean tides—would combine.

But the process of creating life requires more than just sufficient concentrations of the right minerals (chemicals) in a suitable solvent such as water. The problem is that most atoms and molecules lack sufficient energy to bond spontaneously when mixed together. Throwing the minerals into the ocean and stirring them together only leads to well-stirred minerals, not to complex, organic molecules and simple life forms.

Extra energy is needed to combine atoms and inorganic molecules into organic molecules and life. That energy would have two sources on Solon: the sun's ultraviolet radiation and the lightning generated by the storms in Solon's rapidly moving air. Those storms would generate hundreds of thousands, perhaps millions, of lightning bolts each day. Over millions of years the combination and destruction of chemicals would lead to the first life in Solon's oceans.

A key to successful formation, replication, and evolution of life is the presence of large supplies of carbon. Carbon atoms bond with one another as well as with many other elements, providing the essential flexibility needed to create complex life. On Solon, as occurred on Earth, the carbon used to create life in early oceans would come from the atmospheric carbon dioxide that had been dissolved in the water. The carbon dioxide molecule would have been either split into carbon and oxygen or incorporated into other, more complex, molecules. Freed oxygen atoms would have floated upward toward the ocean surface.

Slower Initial Spread of Life on Solon

Once formed, simple aquatic bacteria and algae would spread throughout the oceans of Solon, as they presumably did on Earth. But this initial propagation of life would be much slower on Solon than it was here, due to two problems not encountered on Earth.

First, Solon's lower tides would wash much less sand and soil into the oceans than were brought down to the oceans by tides on Earth. Even the waves created by the strong winds on Solon would not send water as far, or as powerfully, inland as the tidal water flow on the early earth. As we will see in chapter 2, the early tides here were higher, possibly over a hundred times higher, than they are today. Those early waves traveled many miles inland every few hours and brought enormous volumes of dirt back to the oceans. The effects of wind-generated waves a few feet or even a few tens of feet high on Solon would be tiny by comparison.

Second, Solon's lower tides would produce less global ocean movement than there is on Earth. Once the minerals got into the oceans of Solon, the planet's gentle, sun-generated tides would be much less effective at mixing and spreading the chemicals necessary to create life than were the higher, more energetic tides on the young earth. The silt from rivers would therefore remain where it was dumped in the ocean, building up islands at the mouths of the rivers, rather than dispersing and mixing with minerals from other sources as it did on Earth. Since it would take longer to spread and mix the chemicals necessary to create life, the development of early aquatic life would be delayed compared to that on Earth.

Once life is created, it needs to duplicate itself rapidly enough to ensure the survival of its species. In turn, duplication requires that the necessary minerals be sufficiently concentrated so that the existing life can assimilate them. Merely floating in Solon's oceans, simple life would rarely encounter adequate broths of the minerals necessary for replication.

On Earth that problem was overcome. Both organisms and raw minerals adhered to rocks exposed between high and low tides on hundreds of thousands of square miles of beaches. There, along with a choice array of building-block chemicals, these organisms were undisturbed by buffeting ocean currents for several hours at a time. The tiny life forms replicated themselves without the danger that their building blocks would float away. As the tide rose again, the rocks resubmerged and the newly replicated life dispersed. Some organisms moved to new rocks, where the replication cycle was repeated. In this way both the number of life forms and their territories grew.

On Solon, however, the range between high and low tide each day would be much smaller than it was (or is) on Earth. Perhaps a hundred times less shoreline would be uncovered by tidal water, offering far fewer surfaces on which life could reproduce. Once created, the reproduced life forms there would be spread less widely by Solon's feeble tides. This slower initial spread of life on Solon would delay all the later stages of evolution compared with their occurrence on Earth.

Making the Air Breathable: The First Free Oxygen

As noted above, early ocean life on Solon would consume carbon dioxide dissolved in the water and convert it to other carbon compounds and oxygen gas. A waste product of the carbon dioxide conversion process, the oxygen would be released back into the ocean. Some of this oxygen would drift upward and eventually be released into Solon's atmosphere. The same process occurred on the early earth.

Since oxygen is an exceptionally reactive element that combines easily and rapidly with many other elements and compounds, the first oxygen to enter Solon's atmosphere from the oceans would not remain there. During Solon's first 3.5 billion years, virtually all of the freed oxygen would quickly combine with atoms and molecules on the planet's surface and in its air. One of the substances to which it would most commonly bond on the land is iron.

Iron was not always present on Solon's surface, however. When Solon first formed, it would have been entirely molten. Just as dense materials sink in water and light materials rise, the dense iron on Solon's surface would have sunk toward the center of the planet, while lighter elements such as silicon and manganese would have risen. Eventually a small fraction of the sunken iron would have returned to the planet's surface through volcanic activity.

The first free oxygen would quickly join with the exposed iron to form iron oxides, which we call rust. This oxidation also occurred on the earth and on Mars. In fact, the reddish tinge of Mars today is due to iron oxides on its surface. Since the early free oxygen on Solon would be quickly removed from the atmosphere, it would be unavailable for consumption by land animals. This is why early animal evolution would take place in Solon's oceans, just as it did on Earth. From the plant life growing in them, Solon's early oceans would contain the free oxygen necessary for respiration in animals.

Eventually all the elements on Solon's surface and in its air that could combine with oxygen would have done so. Thereafter, the oxygen released into the atmosphere from the oceans would remain and the conversion to a breathable oxygen-nitrogen atmosphere would begin. Earth reached this stage about three quarters of a billion years after it was formed. The process on Solon would take longer, perhaps as much as an extra quarter of a billion years.

Preventing Lethal Levels of Oxygen in the Air

The initial development of surface life on the earth was held back by too little oxygen in the air, and the same would be true on Solon. As on Earth, once atmospheric oxygen became available, exactly the opposite problem would occur; namely, there would be too much oxygen for unprotected life. This is because when it combines with other elements, oxygen gives off energy in the form of heat. For example, mixing hydrogen and oxygen creates water and generates enough heat to propel rockets into space. The growing oxygen content in Earth's early atmosphere provided more heat than the cells of early plants and animals could withstand. Without sufficient safeguards, excessive oxygen would create so much heat as to kill organisms either by disrupting their cells or literally by burning them up. On Earth, early life evolved biological protection against oxygen, allowing it into cells very sparingly. The same would occur on Solon.

Once plant life begins to flourish in Solon's oceans, the amount of oxygen sent into the air would rapidly increase. However, if all the oxygen that was originally part of carbon dioxide molecules in Solon's atmosphere returned to the air, the air would be far too rich in oxygen for even the best-protected animal life to withstand. The oxygen level could be as much as one hundred times higher than it is on Earth. Life as we know it can deal with only about twice as much oxygen as Earth's atmosphere now contains; any more and the cells would overheat and die.

Even if life on Solon somehow developed biological protection against such incredibly dense oxygen, surface life would have absolutely no chance of surviving that atmosphere. The first fire would immediately begin burning everything combustible, creating a fire storm that would annihilate all life on the planet.

Fortunately, as on the earth, only a small fraction of the oxygen from the carbon dioxide in Solon's oceans would return to the air. The oceans would be so massive that more than half of the carbon dioxide deposited in them from the atmosphere would remain dissolved there forever. The remainder would be processed by early sea creatures, who would convert most of it into other carbon compounds. For example, many sea creatures would evolve shells containing carbonate compounds such as calcium carbonate. As the shelled creatures died, they would sink, taking their oxygen-laden carbonate shells with them. Layers upon layers of shells would coat the bottom of Solon's oceans. As the weight of this matter increased, it eventually would compress itself into carbon-rich limestone. The oxygen in the shells would therefore stay out of the air. Similarly formed limestone is found under present and former oceans of the earth.

Some of the oxygen given off by its ocean life would remain dissolved in Solon's oceans. (Similarly, the earth's oceans keep much of the oxygen created by plant life in them; this is the oxygen that fish breathe today.) Happily, only a small portion of the oxygen from Solon's rich carbon dioxide atmosphere would ever see the light of day as free oxygen molecules. In this way the oxygen levels on Solon would be prevented from becoming dangerously rich. As with the earth, Solon's final nitrogen-oxygen atmosphere would have a much lower density than did its carbon dioxide atmosphere.

Although most of the carbon dioxide from the air would dissolve in Solon's early oceans, some of this gas would remain in the air even as the atmosphere is being converted to nitrogen-oxygen. Early plant life on Solon's surface would eventually scavenge this remnant carbon dioxide. Plants use carbon dioxide to create carbohydrates and other biological molecules. In the process of photosynthesis, oxygen is released as an undesirable by-product. These carbon dioxide–processing activities would eventually rid Solon's atmosphere of virtually all its remaining carbon dioxide. This is how the earth's present nitrogen-oxygen atmosphere came into being.

Because it was denser originally, Solon's carbon dioxide atmosphere would be thicker than the corresponding atmosphere of the earth throughout Solon's first 4 billion years of existence. It would therefore take longer—by as much as half a billion years—than it did here to convert all of this gas into free oxygen molecules and other compounds.

Solon's atmosphere began with 15 percent more carbon dioxide than did the earth's. All other things being equal, its atmosphere would eventually contain about 15 percent more oxygen than ours does. Fortunately for the evolution of life there, this increase is in the range within which living tissue could adapt. As we will now see, both the thicker atmosphere and its slower transformation from carbon dioxide to oxygen would lead to differences between life on Solon and life on Earth.

The Extra Carbon Dioxide and the Greenhouse Effect

Throughout Solon's geological history, the higher density of carbon dioxide in the air would keep it hotter than the earth at the same time. This extra heating on Solon would occur because carbon dioxide is a greenhouse gas, meaning it stores more heat than other gases do. Because more water evaporates as the temperature rises, Solon's warmer air would have more moisture in it than the cooler atmosphere of the earth.

Here is a brief explanation of how the greenhouse effect works. Light and heat from the sun enter a planet's atmosphere. Some of this energy is absorbed by greenhouse gases such as carbon dioxide and water vapor in the air. The rest of the light and heat pass to the planet's surface. The heat striking the surface is absorbed, as is some of the light, which is converted into heat by the rocks and water it encounters. This energy from the sun goes into warming the surface. Heat is then radiated back up into the atmosphere. The greenhouse gases in the air absorb some of the heat leaving the surface, thereby warming the atmosphere further. In turn, the warmer atmosphere keeps the planet's surface warmer than it would be without the atmospheric greenhouse gases present.

LIFE ON SOLON

Early Plant Life and the Last Carbon Dioxide

As the carbon dioxide content of Solon's air dwindled, plant life would develop on its surface. Just as they do on Earth, plants on Solon would use sunlight to convert the remaining carbon dioxide into nutrient compounds through photosynthesis. A waste product of photosynthesis, oxygen would be released into the air, joining to the oxygen from the seas.

The atmospheric heat and humidity might suggest that Solon's early plant life would be similar to the luxuriant, often towering growth that occurred on Earth, but this ignores the effects of Solon's winds. These high, persistent winds would inhibit leafy plant growth because wind-whipped leaves would be easily damaged or torn off. Swaying leaves would also work inefficiently in converting carbon dioxide to food because their motion would prevent a uniform flow of fluids through them. Furthermore, they would contribute to the destruction of their parent by catching the wind, allowing it to knock the trees over.

Instead of growing upward and having many layers of

leaves, early plant growth on Solon would grow low to the ground to protect itself against the strong winds. Having fewer layers of leaves, these early plants would have less total leaf area than do plants on Earth.

Less leaf area means slower conversion of the carbon dioxide in Solon's atmosphere into oxygen. As a result, removal of the last remnants of carbon dioxide would be delayed even longer than if the leaves on Solon were similar to those on Earth's early vegetation. Until the extra carbon dioxide is removed, Solon's air would be excessively hot, dense, and still chemically incapable of supporting animal life on land.

Effects of High Winds on Plants

Even squat plants would need secure anchors in the face of Solon's winds. To stay low to the ground, they might evolve frequent and secure roots all along their stems. If necessary, leaves would develop appendages to help them cling to the soil. We see some of these features on the plains and along shorelines of the earth, where persistent winds affect plant and tree life. The arctic willow grows upward only a few inches, while spreading along the ground up to ten feet and more. Such trees are known to live for centuries. The trunks of jack pine and other trees grow bent and twisted under the influence of persistent winds. Instead of growing very high, many trees spread outward, like inverted V's. On Solon, perhaps such arched tree trunks would be the rule rather than the exception.

Eventually adaptations such as deeply burrowing tap roots would enable plant life to grow upward toward Solon's unfriendly but available sky. These roots would enable the trees to withstand the buffeting they would frequently experience. Also, tall trees could evolve shapes that force the wind to push downward on them, keeping them fastened to the ground.

Like the tree itself, leaves would also need greater structural connections than they have on Earth if they are to remain attached to their branches. Otherwise the winds would whip them rapidly back and forth, frequently causing damage. We see this on Earth when long, broad, flexible leaves are buffeted by storms. Wide, flat leaves are more easily damaged by the wind than are short, narrow, stiff ones. This suggests that needles, like those of evergreens, would be favored on Solon over big-leafed palms or elm and maple leaves. This is not to say that pine trees per se would be suitable for Solon. While pine needles would do well in the wind there, the shallow roots of evergreens would make these trees too susceptible to being knocked over.

Once clusters of vertically growing trees became established on Earth, they acted as windbreaks behind which forests or jungles could grow. Since they are protected from wind, the interiors of jungles on Earth are very calm. But on Solon jungle trees would sway more than they do here, since the wind would push harder against their tops. Jungles on Solon would therefore be more challenging places for the evolution of tree-dwelling animals, especially for creatures that leap from tree to moving tree.

Phototropism—Plants Tracking the Sun

Besides the strong wind it creates, Solon's rapid rotation would present another challenge to plant life. Many plants are phototrophic, meaning that they turn their leaves and petals toward the sun to maximize the energy they receive from it. For Earth's heavy plants, such as sunflowers, this already takes a lot of energy.

On Solon, with the sun moving three times faster across the sky than it does here, the energy expended by plants in following it would be enormous by Earth's standards. Phototropic plants on Solon would require much stronger and more responsive internal support structures. If moving so quickly required the use of too much energy, making phototropic plants unviable, more efficient static plant designs might evolve. For example, cylindrical leaves that always have part of their surface facing the sun might prove a successful adaptation.

Early Animal Life on Land

Animal life would also evolve so as to be compatible with its environment. The process of natural selection, in which suitable life forms evolve to inhabit available space, does not follow a preplanned or even repeatable path. In his book *Wonderful Life,* Steven J. Gould points out that there is no guarantee that humans or any other animals that ever existed on Earth would evolve here again even if our planet reran its geological history exactly as it occurred. Evolution is made even less predictable by random events such as massive meteor impacts (which occur on Earth from time to time), storms, and fires. In fact, there is no ideal set of animals to inhabit any planet, including the earth.

At first glance, this insight might suggest that we could propose virtually any creatures we want for Solon consistent with its environment. But real environments are so complex that creating alien creatures out of whole cloth on Solon and the other hypothetical worlds of this book would take us entirely too far into the realm of science fiction. Instead, we will do two things: First, we will consider whether or not there would be biological niches on Solon for animals from Earth. Second, to improve their chances of survival on Solon, we will propose modest variations of them.

Of course, modifying Earth's animals to make them better suited to life on Solon presents new problems. Changes that might seem particularly advantageous from one point of view often turn out to be problematic, if not lethal, from other perspectives. Let us then consider the following example of why structural changes to animals are so perplexing.

Adapting to the Oxygen Concentration

The richer oxygen content of Solon's atmosphere would require biological adaptations in the cells and in the larger structures of animals. In fact, the changes necessary for them to adapt can lead to conflicting physiological structures.

As we discovered earlier, cells on Solon must evolve the ability to regulate the amount of oxygen that enters and leaves them. Cells use oxygen in their metabolic (life) processes. On Earth, the energy from such reactions is used by the cell, while the waste heat generated by metabolism is removed before it can cause any damage. Too much oxygen in a cell creates lethal amounts of heat. Too little oxygen, and the cell would not have enough energy to carry on its functions. Cells on Solon need not use exactly the same amount of oxygen as cells do on Earth. Consider the consequences of cells using more oxygen on Solon.

If animals on Earth are a good example, the extra heat generated by more oxygen in cells would dangerously overheat warm-blooded creatures on Solon. Like their counterparts on Earth, those animals would have internal body temperatures uncomfortably close to the maximum safe level. Warm-blooded creatures are those that keep their internal temperatures constant. Humans, for example, have normal body temperatures right around 99 degrees Fahrenheit, while our brains begin to die when heated to 104 degrees Fahrenheit. Our heat-regulating system is very finely tuned.

Increased metabolism by warm-blooded animals would require even more efficient internal cooling to prevent internal damage. This cooling could occur in one of two ways. Either the extra heat could be used to heat parts of the body that are exposed to extremely cold temperatures, or it could be removed from the body entirely. For example, sea mammals such as otters and seals use some of the heat created by their cells to keep their bodies warm. Their fur and fat help trap the heat of metabolism. (The idea that fur "keeps cold out" is a misconception: Cold doesn't travel; only heat does. Cold is just the lack of heat.) If such animals used more oxygen and generated even more heat on Solon, they would reduce their need for fur coats and thick layers of fat.

Sea mammals could therefore be lighter and more agile on Solon. Reducing fat quantities, however, would actually affect sea mammals adversely in an important but perhaps unexpected way. Besides storing food and providing insulation, the fat of sea mammals adds to their buoyancy. With less fat such animals would be denser than they are now and therefore would sink more readily. They would have to use more energy swimming back to the surface to breathe air. In order to remain near the ocean surface, thinner, hotter sea mammals on Solon would have to have more stored energy (i.e., more fat). The lack of buoyancy would wipe out the advantage of having less fat, namely, being able to swim faster and farther on a smaller supply of fat.

Land animals, if they used more oxygen, would also need to remove the extra heat from their bodies. On Earth body heat is released in animals by air warmed in lungs, by blood removing heat from internal organs to the skin, where it is radiated, and by sweat carrying heat out through the skin. When extra cooling has been needed, large, thin surfaces full of blood vessels have often evolved. The blood flowing through these ears, sails, or other appendages is close to the surface of the skin and therefore easily radiates its heat out of the body.

Large cooling surfaces on Solon's animals would create

problems, however. The planet's strong winds would push on them and destabilize the animal. While the massive dinosaur dimetrodon, with its towering cooling sail, might possibly be able to withstand the winds on Solon, lighter animals with cooling sails or fins or even especially large ears like those of hares or foxes would not do well there.

Altering animal designs to better fit Solon is clearly complicated by the multiple uses of many body parts. While this realization will not deter me, the reader is urged to keep the modifications suggested hereafter in that perspective. We turn now to explore some of the differences between the biological niches on Solon and those on Earth.

BIOLOGICAL NICHES

Successful plant and animal species have biological structures that make them suitable for surviving in a certain environment, called a biological niche. On Earth, virtually every surface, body of water, and part of the air contains numerous different forms of animal life. I once took my oldest son rock hunting, and after splitting one rock open, we found a dozen tiny, red insects crawling around inside it. There seems to be no end to nature's ingenuity in adapting life to the space available.

Because of the differences in tides and winds, however, there would be fewer biological niches and less diversity on Solon than there are on Earth. We will explore some of these niches in detail, starting with the beaches.

Effects of Lower Tides on Sea Life

Solon's lower tides would diminish the intertidal beach zone, the area of beach periodically covered by water. On Earth the intertidal zone of nearly every beach is brimming with different life forms. Indeed, it is a specialized niche, occupied by life forms that can take advantage of both the water and the dry land.

Buried along Earth's shores in the intertidal mud and sand are hundreds of different types of shelled animals and insects. Nature ensures survival of such species by encouraging them to proliferate. Habitats must be large enough for sufficient members of a species living in them to withstand disease, extremes of weather, and attack by predators.

Solon's lower tides would narrow the intertidal beach area, thereby making it harder for many species to maintain safe populations. This increases the likelihood that competition or disease would wipe out entire colonies of tidal animals. The diversity of life in the narrower tidal zones on Solon would be more limited than on Earth so that species evolving in them would have enough room to survive.

The change in the heights of tides on Earth throughout a month has affected the life cycles of some sea creatures. For example, grunions spawn ashore only at the highest reaches of the highest tide, called spring tides. (The name "spring" refers to the water "springing up," not to the season of the year.) Spring tides occur when the moon, sun, and Earth are in a straight line. At that time the moon and sun work together to generate especially high tides; this configuration produces the greatest distance between consecutive high and low tides.

Grunions are carried far up on the beach during the spring tides. The fertilized eggs they deposit there remain safe from scavenging denizens of the sea. Since tides over the next two weeks would be lower, other fish and crustaceans would be unable to get to the eggs without leaving the water.

Since all the tides generated by the sun alone would be equally high, Solon would lack spring tides. As a result, grunions or fish like them would be prevented from safely depositing their eggs far up the beach. Spawned at the standard high tide, grunion eggs would be susceptible to being washed away or consumed by fish carried up to them on the next tide. Because tidal cycles would change little throughout the year, life in Solon's intertidal zones would have greater stability than the equivalent life on Earth.

Effects of Lower Tides on Birds

The decreased diversity of sea creatures in Solon's intertidal zone would have its effects on birds that feed on creatures in wet sand and on beached seaweed. Competition would be fiercer among different species of tidal feeders. The birds equivalent to our curlews, godwits, gulls, herons, jacana, oyster catchers, plovers, sandpipers, and snipes would have less food available to them. Indeed, on Solon the lack of space in which to feed would prevent many of those species from evolving at all.

The diminished tidal habitat on Solon means that there would be more area along the shore where shallow water remains throughout the day. This would increase the feeding grounds for diving birds such as our loons, grebes, cormorants, gannets, and auks. This shows that while the diversity in niches would decrease on Solon, it seems likely that the same physical space would remain filled with life.

Effects of High Winds on Animals

Emerging from the relative safety of the oceans onto the windswept beaches of Solon, animal life would encounter more difficulties than did similar life on Earth. First of all, the more rapidly moving air on Solon would make respiration more difficult. You can experience the same effect today by walking into a strong wind; it becomes very hard to breathe. Strong muscles for breathing and well-shielded noses would be useful adaptations on Solon.

Besides making breathing difficult, strong winds remove both heat and moisture from the bodies of land animals. These problems would be worse on windy Solon than they are on our calmer Earth. The greater heat removal would actually help on Solon if animals there have higher metabolism rates, as discussed earlier.

This extra energy available from increased metabolism would help some animals to compensate for difficulties in traveling created by Solon's turbulent weather. Consider birds. Today they are lightweight, hollow-boned fliers. They fly well either with a good tail wind or in calm air, but they are handicapped by strong head winds and rough air. Most of today's birds would be unable to cope with Solon's strong, persistent winds.

However, sturdier and stronger fliers could evolve on Solon by developing stronger muscles and using the extra energy provided by the planet's oxygen-rich air. Such birds would be better suited to life in the tumultuous skies of Solon than would Earth's relatively delicate aviators.

On land, the abilities to remain upright and to move without being pushed or rolled over by the wind would also be early challenges for animals on Solon. Long claws that dig deeply into the soil would help animals stay upright in the face of strong winds. So would wide, low bodies. Animals might even use the wind to keep themselves in place by evolving aerodynamic shapes similar to that of the spoilers that help hold race cars on the track.

Solon's high winds would carry pieces of rock, massive hailstones, and abundant quantities of sand and topsoil over long distances. Because both animal and plant life would have to withstand impacts from this debris, possessing a thick skin would be beneficial. Hard-shelled creatures such as tortoises and armadillos would be particularly suitable for this environment, since they can shield their soft body parts during especially dangerous times. Ponderous creatures such as dinosaurs would probably also thrive in Solon's more windy environs, while the smaller, lighter mammals that coexisted with them on Earth would have more difficulty there.

Biological Clocks

The most profound effect of Solon's rapid rotation comes not from the wind it generates but from the shorter, faster day, which would dramatically change the rhythms of life. Animals and plants on Earth are regulated by built-in timing mechanisms called biological clocks or circadian (almost daily) rhythms. These clocks drive daily and seasonal activities such as waking, sleeping, resting, food gathering, eating, and mating.

The dominant clock in most things living on Earth today has a natural rate that is approximately, but rarely exactly, equal to twenty-four hours. Many animals have clocks with twenty-three-hour cycles, while humans have twenty-five-hour cycles.* To stay in sync with the twenty-four-hour day, circadian rhythms need to be reset daily. Fortunately, the natural biological clock rate is easily overridden and synchronized to the day by external stimuli, especially the cycle of day and night. Other external effects that can reset biological clocks include changes in temperature, air pressure, sound, and food and water availability. This process of resetting biological clocks is called entrainment.

Without their biological clocks being entrained, animals would quickly go out of sync with their environment. They would begin sleeping when they should be eating, hunting when they should be sleeping, and so on. It is not surprising, then, that most biological clocks can be entrained by external cycles.

*Natural-clock rates are determined experimentally by isolating test animals and humans. Without the benefit of changes in light and dark, temperature, availability of food and water, social cues, air pressure, or sound, the animal lets its natural-clock rate take over; it begins functioning at that rate. Instead of sleeping in a twenty-four-hour cycle, it would sleep in a cycle determined by its circadian rhythm.

Some people have biological clocks that aren't entrained by external stimuli. These people have difficulty functioning in society. Their waking and sleeping patterns cycle through a twenty-five hour day, often making them dysfunctional during the daylight hours. Humans traveling by jet across time zones experience the same problem, namely, that their circadian rhythms need to be re-entrained after they land. Readjustment from so-called jet lag can take several days.

Biological clocks can be entrained only by external cycles that differ from their natural rates by less than three hours. In other words, a plant or animal can be entrained by the earth's day-night cycle only if its natural circadian rhythm is between twenty-one and twenty-seven hours long. Otherwise, the plant or animal lives its life based on its circadian rhythm, usually with disastrous results. No Earth plants or animals could function for long on Solon, with its eight-hour days. The days on Solon would be too short to entrain a twenty-five-hour circadian clock; humans from Earth, if transplanted to Solon, would be doomed to eternal jet lag.

Indeed, when humans on Earth eventually do migrate to the stars, one of the innumerable things they will have to know about their destination planet is its rotation rate. If the planet's day is not between twenty-two and twenty-eight hours long, the people arriving there are going to have enormous difficulties adapting to its rhythm of life.

All living things on Solon would evolve much shorter biological clock cycles than those that occur on Earth. The biological clocks would have to run within a few hours of Solon's eight-hour days. What toll would this rapid cycling of activities take on Solon's creatures? Could animals do all that they need to in such short days?

Life Cycles

Most animals on Earth need more than four or five hours of daylight to complete even such fundamental tasks as migrating, foraging, eating, sleeping, mating, building, teaching their young, and watching for predators. Therefore, Solon's shorter days would require adaptations to allow creatures there to accomplish the essential activities of life. One possibility is for activities that are completed in a day on Earth to be spread out over several days on Solon. Consider the activities involved in gathering, preparing, and eating food.

Ideally, most warm-blooded species on Earth eat several times a day, although almost all can survive on single feedings if necessary. On Solon it would seem essential that all species function on one meal a day or perhaps even one meal every two days. Consider the somewhat special case of the first people who evolve on Solon. Like those of our ancestors, their lives would be organized around reliable food supplies.

Both the kinds of food (meat versus vegetable) and the ways they are consumed (rapidly or leisurely) would be affected by the length of the day. Because Solon would have less daylight and darker nights, hunting would be harder for those early people. We know that the process of tracking, killing, and retrieving game often took our ancestors on Earth more than four or five hours; these activities would take more than an entire day on Solon. Early hunters there would need more "days" to secure their quarry than did Earth's early hunters, especially when large numbers of people need to be fed.

These difficulties in "daily" food gathering could well motivate the discovery and implementation of agriculture for sustenance much earlier on Solon than it happened here, which was about ten thousand years ago. By farming, tribes could stay in one place longer; time can be used more efficiently growing and gathering food than following migrating animal herds. Essentially all community members could productively participate in farming, whereas entire hunting parties could come back empty-handed, their time wasted.

The more concentrated oxygen in Solon's atmosphere might enable all life, including large creatures like ourselves, to do things more rapidly than on Earth. Animals could travel, hunt, eat, fight, mate, and even think faster. Life on Solon might run like a videotape on fast forward.

The difference in length of days also raises the question of whether the number of days an animal lives affects its total life span. Since Solon would orbit the sun once a year, just like the earth, would creatures on Solon live the same number of years as creatures on Earth? Or would the greater number of days each year affect the life expectancy on Solon? We cannot answer these questions.

Menstrual Cycles

While the moon's influence on the earth's rotation rate directly affected the biology of circadian rhythms, other perceived effects of the moon on humans are far more questionable. The best example of this is the female menstrual cycle. The human female menstrual cycle, which prepares the uterus for nourishing a fertilized egg, occurs every twenty-eight days. It is therefore close to the lunar cycle of twenty-nine and a half days, leading many people to infer that they are related.

Although some studies of the onset of menses argue for a correlation with the phases of the moon, most research finds no correlation. From another point of view, the alleged cause-and-effect relationship between the moon's phases and the menstrual cycle is even more suspect. Implicit in the link between lunar phases and human menstruation is the assumption that other animals, living under the same moon, would have the same menstrual cycle. This is not the case.

Very few animals have a fertility cycle even remotely similar to the twenty-nine-and-a-half-day cycle of lunar phases. Chimpanzees have thirty-seven-day cycles, whereas sheep have eleven-day cycles. It seems likely, then, that menstrual cycles of varying lengths would also occur on Solon and that human menstrual cycles might still be twenty-eight days long. Conversely, if the human menstrual cycle is somehow entrained by the moon, then the absence of the moon would set it free to run at its "natural" rate with no harm done.

Development of the Senses

Since the lives of animals on Solon would be crammed with activities essential for survival, the higher winds and shorter days there would put great demands on their senses. The senses are passive biological structures designed to receive external stimuli. Our taste buds, noses, eyes, ears, and skin do not generate tastes, smells, sights, sounds, or pressures; they only receive sensory information from outside our bodies. Each sense provides animals with information about threats and opportunities occurring at different distances from their bodies. Taste tells about matter entering the body. Touch tells about physical contact. Smell tells about nearby matter. Hearing tells about things within a hundred feet or so. Sight is special in that it provides information from all distances from inches to billions of miles.

On Earth, animals with an underused sense usually have an underdeveloped sense organ. For example, many fish living in the deep oceans and many animals living underground lack operational eyes. Conversely, senses that are used frequently or under adverse conditions, such as the eyes of owls, become highly refined.

Besides the traditional five senses, at least one more passive sense began to evolve on Earth. This is the ability to detect the presence of heat or infrared radiation, a form of energy similar to visible light. Heat sensing became specialized to keep animals from burning themselves. In most species these sensors did not evolve to the state where an animal could do more with them. We will explore enhanced heat-detection in detail in chapter 9.

Because the sensory equipment of creatures on Earth is passive—that is, the body does not have to generate the source of taste, smell, light, or sound—these senses were relatively easy to evolve compared with senses that would require animals to emit as well as receive sensory energy.

An apparent objection to the concept that animals' senses are passive is the ability of animals to make sounds. However, in understanding hearing, it is important to separate the ears and the voice apparatus. In the first place, the two are quite separate structures, each of which could work without the other. Furthermore, ears would still be useful even if all animals were mute. In that case, ears would help in locating prey and predators via their footfalls and the sounds they make moving through vegetation. Ears also would warn of severe weather activity, water flow, and other natural phenomena.

To be sure, ears became more useful when animals developed the ability to create sounds. Then members of a species could locate and warn one another at a distance. They could also scare away enemies with their voices. Eventually, the human ability to communicate complex thoughts with sounds enabled our ancestors to begin to develop social structures.

Senses that do require their owners to actively emit as well as receive sensory energy have actually evolved on Earth to aid certain species. Sonar in bats is a good example. These animals emit high-pitched sound and then hear echoes that have bounced off nearby objects. A bat's ability to hear such echoes would be useless if the animal did not emit specialized sound energy in the first place.

On Earth, in most cases the evolutionary experimentation necessary to create active senses was apparently far too expensive for nature to undertake, or at least to complete by now. Only when it was essential for survival, as in the case of the nocturnal bats who live in pitch-black caves and hunt at night, did nature expend the full measure of energy necessary to perfect them. The noisy, blustery conditions on Solon, however, might be sufficiently challenging to the survival of complex life forms so that such senses would be indispensable. Let us now consider how hearing and sight might evolve on Solon.

Hearing

The strong winds on Solon would frequently mask the kinds of sounds that can be heard on Earth, including many animal sounds. It would also prevent the sounds from traveling well upwind. On Solon, to detect nearby animals through the sounds they make would require improvements over the hearing apparatus of many animals, especially humans, on Earth.

Like those of deer, horses, rhinoceroses, and other animals on Earth, directionally sensitive ears would help reduce the background noise on Solon. They would do so by preventing the ear from hearing sounds not coming from the area of interest. Reducing background noise serves to increase the sound associated with the object of attention.

This technique has also been applied by humans. Ornithologists interested in the sounds of distant birds often use parabola-shaped reflectors as specialized ears. These large reflectors collect great amounts of sound from the bird, of course, but the reflector's shape also keeps out sounds from other directions. Many animals, to compensate for the greater directionality of their ears, have evolved the ability to swivel their ears easily in almost any direction. Like other animals, people on Solon might have cone-shaped ears to scan around their heads for important sounds.

A second evolutionary improvement to hearing on Solon relates to sound processing in the brain. Perhaps the persistent sound of the wind could actually be filtered out in the brain. This would help clarify and enhance the distinctive noises of animals. On Earth engineers use this principle of filtering unwanted but well-characterized sounds in the noise-reduction circuits of cassette-tape players.

Not only could ears and brains be refined for Solon, but desirable sounds, such as mating calls, could also be adjusted to travel farther in noisier environments. One obvious way to send sound farther is to increase its volume. But at least as important as loudness are the pitches (frequencies) that make up the sound. Different frequencies travel better through windy air than through calm air. Distinctive mating calls would evolve using sounds that carry far through Solon's air. Such calls would enable members of the same species to hear one another over the greatest distances, giving them the best chance of locating one another.

The sense of hearing would probably remain useful on Solon for the general protective purposes sketched thus far. However, the extra noise from the wind might make more subtle uses of sound, such as complex human communications, less practical. Advanced communication skills among Solon's people would be essential, and so other forms of intercourse would have to evolve to replace oral speech.

Sight

Visual "speech" on Solon would be an effective alternative to oral communication. Hearing-impaired people make use of this ability today in the form of sign language. Natural sign language might also evolve with the development of moveable appendages. These limbs might be biological semaphore flags analogous to those found on some ships for communication with nearby vessels.

People on Solon could develop languages to be used by their semaphore limbs. Consisting of visual patterns, these languages would visually present information we now express orally. The semaphore limbs could be effective communications devices, creating fast-changing patterns without expending excessive energy. Perhaps the limbs would contain a variety of "branches" moving independently to express different parts of speech or sophisticated nuances. However, these limbs would have two serious limitations: They would only be effective during daylight hours, and they would have to be shielded from buffeting by the wind.

Biologically generated light that could change intensity or color would be an effective means of communication on Solon. Biological light sources exist in profusion on Earth. Among the generators of such energy are some bacteria, fireflies, krill, jellyfish, flashlight fish, and lantern fish. Squid communicate with one another by changing color patterns over their bodies. If people on Solon inherit such an adaptation, their kaleidoscopes of changing colors and changing light intensities could eventually take on the meanings of words.

Fairly complex chemical processes are required to generate varying light intensities and colors biologically. This has limited the evolution of active light sources in animals on Earth. But in Solon's inclement environment, the effort would pay off handsomely. The ability to communicate using light has three distinct advantages over semaphore communications. First, the light source would not have to be a mechanical limb subject to the effects of winds and rain. Second, light would enable animals to communicate at night. And third, animals or people creating their own light would have the valuable secondary advantage of using it to see at night.

Animals at the top of the food chain would benefit enormously by using biological light sources to help them locate their prey on Solon. Without fear of being attacked, these animals could use their lights whenever they chose.

This would give them an enormous advantage over creatures lower on the food chain whose lights, if they had them, would act as beacons to animals that hunt them. Imagine a tyrannosaurus rex, lights glowing, searching for a choice stegosaurus to eat.

Our present ability to detect visible light has focused our attention so far on this very limited part of the electromagnetic spectrum. And yet nature could evolve communications in other parts of the spectrum, especially using radio waves. Besides visible light, the electromagnetic spectrum also contains radio waves, infrared and ultraviolet radiations, X rays, and gamma rays. (Sound is not electromagnetic in nature. Rather, it comes from the compression of the air between the source and the hearer.) Considering the awesome array of radio-communications devices we humans have developed just over the past century, it seems plausible that nature could evolve radio communications over several hundred million years. If the ability to send and receive radio messages evolved in animals on Solon, we would consider them to be telepathic.

A New Sense: Telepathy

The evolution of large, stable social groups of humans on Earth required the development of exceptional communications abilities. Most of the interactions between early peoples were oral. Our ancestors verbally warned children of danger, dealt with enemies and allies, shared feelings with friends and lovers, and transmitted clan lore, as well as using speech to help carry out all the other interpersonal activities essential for human survival. Only in the past five hundred years have printed words been readily available, shifting the focus of communication from sound to sight.

As we've seen, oral communication was possible on Earth because our planet is such an amenable place for life. Life in general, and oral communication in particular, would be more difficult on Solon. During the hundreds of millions of years just prior to the evolution of people on Solon, ancestral animal species would struggle merely to survive on the planet's more hostile surface. Their limited ability to communicate orally would act as an impetus for nature to explore alternative communications systems.

Semaphore limbs might do well for animals living on the ground in relatively calm, protected areas. Biologically generated light would be especially helpful for nocturnal creatures. However, animals living in trees or in other challenging and noisy environments would need a way to communicate that complemented their complex travel, feeding, and fighting activities, all of which depend heavily on visual ability. For example, primates who jump from tree to tree must concentrate their vision on the process of getting around. They don't have the time to look for lights flashing or semaphore flags waving from other directions. Indeed, living in dense foliage would limit the effectiveness of these visual communications devices.

With sound more difficult to process and sight focused on other vital matters, telepathy would be extremely useful. Because this ability or sense would be completely independent of the other senses, it would supplement rather than interfere with other activities. Also, telepathy could be entirely internal, thereby preventing an external organ, such as an ear, eye, or mouth, from being exposed to the elements on blustery Solon.

At best, telepathy remains underdeveloped in the vast majority of people on Earth. Indeed, the purported ability of some people to "hear" or sense the thoughts of others could be the result of rudimentary radio reception in their brains. Telepathy probably did not develop any further here because the number of evolutionary steps needed to develop it was greater than the number of steps required to develop passive senses. Also, the other senses have proven adequate for animal needs on Earth. But the potential for telepathy does exist.

Our brains give off weak but measurable radio waves. These signals are generated as a by-product of the brain's normal functioning. To be useful for interpersonal communication on Solon, radio "brain waves" would need to be more powerful and to be modulated or changed by the equivalent of a voice box. Combined with the evolution of biological radio receivers analogous to our eyes, this controllable, more powerful transmission would lead to telepathy.

Once developed, biological radio transmitters would also be available for helping animals see in the dark. This would be especially helpful on Solon, since moonlight would be unavailable. Like sonar (high-pitched sound waves) emitted by bats, radio waves could bounce off some objects and help nocturnal animals "see" at night. As an example of using radio waves to visualize otherwise invisible objects, consider that astronomers use this technique today to see in exquisite detail the cloud-covered surface of Venus.

Loss of Moonlight

There would be no moonlight on Solon. Hunting, foraging, mating, and other nocturnal activities there would have to occur without it. This change would affect both predators and their prey. They would have a harder time detecting each other and would have to rely on other senses to do so. On Earth, certain fish, such as the moon wrasse, depend on moonlight for illuminating food at night. Animals that require moonlight to survive on Earth would never evolve on Solon.

One of the more interesting roles of moonlight in biological activities on Earth is its use in initiating the change of some freshwater fish into saltwater species. Consider salmon, for example. These fish are born in freshwater rivers and streams, with their cells designed to function in low-salinity (fresh) water. They swim to the oceans to live, eventually returning to their native rivers to spawn. If the salmon swam into the salt water of the ocean without changing their cell structures to accommodate this different chemistry, they would quickly die from excessive salinity.

Upon reaching their river's mouth, salmon chemistries are transformed to work in the high-salinity (salty) ocean water. The catch is that this change is initiated by the phase of the moon. It occurs only when the moon is new (a sliver). Since this lunar stimulus would be unavailable on Solon, salmonlike fish, if present, would have to initiate their biological change some other way.

If Not from Tree Dwellers . . .

The presumption that humans would evolve from tree-living primates on Solon as they did on Earth carries with it the assumption that suitable trees would exist there. A further assumption is that if such trees do exist, the high winds and the detritus they carry would not prevent them from being a suitable niche for primates. If people don't evolve from tree dwellers on Solon, would that mean that sentient life would never exist there? Not at all. Several routes to the same evolutionary destination often exist. On Earth bats fly, even though their wings evolved through an entirely different path than did bird wings. Pandas have thumbs that evolved differently than the thumbs of primates.

In the same spirit, species other than tree dwellers on Solon might well need the same mental and mechanical apparatus that we humans inherited from our arboreal ancestors. For example, ancestors of humans on Solon might be animals living in regions of great seismic activity, such as the boundaries between tectonic plates. Facing frequent, unpredictable, and potentially lethal dangers, they would need to evolve the capacity to respond quickly and creatively to their hazardous environment. Since their evolutionary needs would be quite similar to the needs of tree dwellers on Earth, they might evolve opposable thumbs and sufficiently complex brains to become aware of their own existence. Evolution is a process that fills voids and expands capabilities in whatever animals are available.

The Loss of the Moon As a Clock

On Solon early humans, facing the challenge of gathering enough food for large groups of people, would encounter greater hardship in developing farming communities than occurred on Earth. On Earth the moon served early humans as a clock, ticking through its phases. This cycle was especially important to the development of agriculture. The moon filled the gap between two other, very disparate natural clocks: a day and a year. The moon's easily identified twenty-nine-and-a-half-day cycle of phases provided valuable, easily remembered intervals for planting and harvesting. The lunar month (the cycle of lunar phases) was also split into smaller units by those who accurately observed the moon's phases.

In the millennia before reliable written records, it was much easier to keep track of lunar months than it was to keep track of individual days. Nowhere was this more important than in agriculture. As early populations of humans grew, organizing large groups of people to follow migrating game became increasingly difficult. Therefore, farming became increasingly important for the reliable provision of food. The demand for dependable agricultural food supplies required great care on the part of farmers. When they planted too early, their crops were killed by frost in the spring. When they planted too late, their immature crops were killed by frost in the fall.

For example, farmers in one region knew that the earliest they could plant crops safely was five lunar months after the day the noontime sun was lowest in the sky. Keeping track of five lunar months was easy, even in preliterate times, when it would have been very hard to remember and tabulate 148 days to the planting date.

The moon was also used as a clock for other purposes. Migrating people, such as many Native American tribes, used the moon to keep track of how long they had been traveling as well as to determine when they should move again. To be sure, a year on Earth is not a whole number of lunar months. There are approximately twelve and a third lunar months each year. As early calendar makers learned the hard way, cycles of lunar phases are unreliable for determining the length of an entire year. Rather, lunar months are best used to measure time from a day established by other means, such as the day of the year when the sun is lowest in the sky (December 21, the winter solstice and the day with the least sunlight).

How, then, might early people on Solon manage without the monthly lunar calendar? Unfortunately, there is no simple alternative either on the planet or in the sky. Natural events on Earth, such as volcanic eruptions, are irregular, nor is the weather sufficiently reliable to use as a clock. Motions of planets through the sky are slow; moreover, the planets are not in the same places among the stars every year. It is much harder to use the other planets as timekeepers than it is to use the moon. Using the positions of the planets among the stars to determine planting seasons requires a successful theory of the heavens, called a cosmology, to reliably predict the positions of the planets and stars in the night sky.

In a practical sense it doesn't matter what cosmology early people on Solon adopt to help them keep track of time. It could be Solon-centered or sun-centered, just as long as it works. On Earth a practical cosmology was not devised until tens of thousands of years after the advent of the earliest humans. It was not until the fourth century B.C. that the Greek astronomer Eudoxus developed a useable model of the universe. Although Eudoxus's Earth-centered theory was physically wrong, it predicted the positions of the planets among the stars with enough accuracy to be useful to farmers and others. Without the moon to help the farmers of Solon, a practical astronomical theory would be needed there thousands of years earlier.

Moonless Astronomy

While the moon aided early people on Earth in keeping track of the seasons, it eventually became a nuisance to astronomers who wished to locate and understand the distant stars and galaxies. Because of the vast distances between our solar system and the stars and galaxies visible from here, these objects are extremely dim compared to the sun or even to the sunlight reflected off the moon. This "moonlight" brightens the earth's atmosphere at night, washing out light from all but the brightest stars and seriously limiting astronomical observation from Earth.

Facing a truly dark sky each night, astronomers on Solon would be able to make observations of dim objects in space every clear night. Above Solon's incessant storms, astronomers on mountaintop observatories would be able to collect much more information about the universe than we acquire from our telescopes on Earth.

On the other hand, the presence of the moon will help future astronomers on Earth make even better observations than astronomers would ever make from Solon. Observatories built on the moon's far side (which never faces the earth) will be able to make observations around the clock with much greater clarity than any made on Earth.

Astronomical observations made from the moon will be sharper than observations made through Solon's atmosphere because the moon has no air. Air acts like a lens to starlight passing through it. Every second the lens of air between a star and the earth (or Solon) shifts as the air's density changes. This causes starlight to change direction as it passes through the air. As a result, the stars appear to change their location and brightness in our sky. These same changes, called twinkling, smear out images of stars on photographs taken through any atmosphere. Therefore all pictures of stars taken from the surface of Solon would be blurred compared with pictures taken from the airless moon.

The Moon As a Goal

Besides its potential as an atmosphere-free observatory, the moon will serve as a convenient stepping stone for human colonization of the solar system. The first step in that direction took place in 1961, when President Kennedy set a ten-year goal of landing humans on the moon. While ambitious, that plan was economically and technologically reasonable. Indeed, the moon is close enough so that we were able to reach it with only thirty years of rocketry and computer science behind us.

A leader on Solon who wished to establish a major human presence in space would have only two choices: Send astronauts to Mars or build a space station. Both of these would be exceptionally challenging for emerging space technologies. Even the gung-ho, can-do rocket scientists of the 1960s knew that Mars would have to wait. We lacked the equipment to transport people there safely, much less return them to Earth.

Solon's early space scientists and engineers could probably design and build a crude space station. However, as we learned with our first station, Skylab, launched in 1973, space stations lack the romance of visits to other astronomical bodies. Used only briefly, Skylab was abandoned to a fiery death in 1979 as interest and funding waned. It was eventually allowed to burn up re-entering the earth's atmosphere. Had there been sufficient public interest in it, Skylab could have been kept in space. Human counterparts on Solon might also quickly lose interest in funding such a project unless the facility was to be the departure point for other space activities such as exploring Mars.

Because of the capabilities we gained from the Apollo moon missions, we have already developed and sent spacecraft to all the planets (except Pluto), as well as to some comets and asteroids. Our scientists accomplished this within thirty-five years of the placement of the first satellite, *Sputnik,* into Earth orbit. Without the moon beckoning to the people of Solon, space technology would probably advance there much more slowly than it did on Earth.

CROSSROADS

Removing the moon seems harmless enough at first. Of course, Solon would differ from the earth. The tides would be lower without the moon, and it would lack eclipses and romantic, moonlit nights, but in the global scheme of things these changes seem trivial. As we dig deeper, we discover that lower tides, higher winds, and shorter days would greatly affect Solon's geography, its ability to evolve life, and the quality of the life animals would have there. As the differences between Earth and Solon become more evident, it becomes clear that Solon would be a much less hospitable place in which to live.

There is much more that could be said about Solon, but this chapter raises a broader question that cries out for consideration. That is, just how ideal a planet did we inherit compared with the one we might have gotten? Are we living on the best of all possible worlds? If not, what would different astronomical conditions in the solar system lead to?

Motivated by a desire to understand the quality of the earth as a habitat, we now explore other worlds which, like Solon, are identical to the earth in all but one or another astronomical condition. Reversing the cornerstone of this chapter, we next consider what Earth would be like if our present moon was much closer to us than it is today.

WHAT IF THE MOON WERE CLOSER TO THE EARTH?
LUNHOLM

ANXIETY IS THE DARK SIDE OF CHANGE. EVEN A THING AS SEEM-ingly benign and repetitive as the sun's motion across the sky disturbed early humans. Each day the sun sweeps across the sky. But where it rises in the east, how high in the sky it gets, where it sets in the west, and how long it is up change daily. As if they didn't have enough to worry about with the challenges of finding food and withstanding onslaughts of disease and inclement weather, early people were unsure if the sun would rise every day. Worse still, as autumn turned into winter, they saw the sun rise farther south and the number of daylight hours dwindle. Would this shortening of the day and cooling of the weather end, they wondered? Would the sun begin to rise higher and stay up longer again? Or would it just stop rising alto-gether, leaving them to freeze to death?

Such fears about the world were bred from ignorance about the physical relationship between the sun and the earth. The anxiety only began to abate during the fourth century B.C. It was then that the Greek astronomer Eudoxus proposed the previously mentioned Earth-centered theory of the universe that seemed to explain the sun's behavior. By explaining why the sun rose and set and why it changed height in the sky throughout the seasons, the theory provided assurance that the sun would indeed rise higher in the sky again after months of getting lower each day.

The masses came to accept the assurances of the philosopher-scientists of the time that the cycle of seasons would repeat each year. But discrepancies between the predictions of the Earth-centered theory of the universe and observations of the sun, planets, and stars arose almost immediately, causing concern to the philosopher-scientists themselves.*

But the Earth-centered theory satisfied too many human needs to be easily overthrown. For two millennia, philosopher-scientists were willing to tolerate inaccurate predictions from the Earth-centered theory rather than face the anxiety of throwing it out and finding a fundamentally different theory that worked better. The Earth-centered theory was patched up with elaborate, physically unjustifiable refinements. Besides working tolerably well, the theory had the added advantage of having the Earth at the center of everything, which neatly matched prevailing theologies.

Two thousand years later the anxiety of change returned

---

*We consider the scientists of the Greek era to be philosopher-scientists because they possessed neither the physical understanding of nature nor the mathematics necessary to describe its phenomena quantitatively. The physical ideas put forward before the sixteenth century A.D. were presented as philosophies, in present terminology, rather than scientific theories. These ideas were qualitative, difficult to test, and often open to differing interpretations. With the advent of Sir Isaac Newton's calculus and the mathematical description of physical phenomena in 1687, scientific theories became more concrete, more easily tested, and more specific in their meanings and predictions. From then on philosophy and science became separate disciplines.

to astronomy. This time it was not due to fear that the sun would vanish during winter, but to the evidence that the Earth-centered theory of the universe was undeniably and irreparably wrong. The more precise observations of planets and stars that came from astronomers of the day could not be reconciled with the theory. In 1543 Nicholas Copernicus proposed the replacement theory in his book *De Revolutionibus Orbium Celestium* ("On the Revolutions of the Celestial Spheres"). Copernicus claimed that the earth and other planets orbit the sun and that the earth spins on its axis. The earth's orbit around the sun creates the seasons, while the spin of the axis creates day and night.

This sun-centered cosmology made more accurate predictions than the Earth-centered cosmology it replaced. However, the physical explanation for how the sun keeps the planets in orbit had to wait until 1687 when Sir Isaac Newton published his *Philosophiae Naturalis Principia Mathematica* ("The Mathematical Principles of Natural Philosophy"). In *Principia* Newton explained the principle of gravitation, the force that keeps planets in orbit around the sun and moons in orbit around the planets. At last people had the means to understand physically (rather than philosophically or theologically) the motions and positions of the planets and stars. While the scientists' anxiety dissipated after Newton explained the law of gravity, the removal of the earth as the center of the universe continued to cause grave concern in religious circles for centuries.

The scientific relief at understanding the gravitational relationship between the earth and sun (and similarly between the earth and moon) lasted less than two centuries after Newton's day. During that brief period, people were quite content to believe that the earth's land, oceans, weather, tides, and seasons had been constant throughout the planet's history, and that they would remain so forever. They couldn't have been more wrong.

The earth is continually undergoing both cyclic and irreversible changes. For example, the seasons reverse in a cycle lasting twenty-six thousand years; thirteen thousand years from now January will be the height of summer in the Northern Hemisphere. The earth also periodically plunges into ice ages lasting tens of thousands of years. Among the irreversible changes, many are caused by tectonic-plate motion: Some continents move apart, others collide, they all change shape, and oceans widen or shrink. The slowing down of the earth's rotation rate causes the irreversible change of the days growing longer. Indeed, every object in the solar system changes as it ages.

Irreversible evolutionary change in the solar system was first proposed by Sir George Darwin, son of evolutionist Charles Darwin. Among other things, Sir George proposed that the moon is now and always has been spiraling away from the earth. The mechanism for this will be described shortly. Although Sir George published this theory in 1897, it was not until 1969 that high-tech observations of the moon proved him correct.

In November 1969, *Apollo 12* astronauts Charles Conrad, Jr., and Alan Bean set an elegantly simple experiment on the moon's surface. The apparatus consisted of an array of reflectors similar to the red and orange ones that make cars more visible at night. Such reflectors return light that strikes them in exactly the direction from which it came. The astronauts aimed the array in the general direction of the earth and left it there upon their return home.

Powerful laser beams were then shot from Earth toward the reflectors on the moon. Extremely accurate measurements were made of the time it took for the pulses of laser light to travel to the moon, bounce off the reflectors, and return to Earth. Astronomers then calculated the distance to the moon from the speed of light (186,000 miles per second) and the travel time. The distance to the moon is merely the speed of light multiplied by half the round-trip travel time. These measurements are accurate to within a few inches. What is more, repeated over a period of years, these laser-ranging experiments have shown that the moon is moving away from the earth at a rate of about two inches per year.

The moon's irreversible movement away from Earth is the basis of this chapter's scenario. What would the earth and life on it be like if the moon were closer today? This question is fundamentally different from the one we pursued in the last chapter not only because of the new phenomena that result if the moon is closer, but also because of the different physics we must consider in relocating the moon.

In chapter 1, preventing the moon from forming required only that the Mars-sized planetesimal that splashed the moon into life travel in a slightly different orbit and thereby miss the earth. The change in that planetesimal's orbit presents no physical or conceptual problem. In fact, since Mercury and Venus have no moons and since the tiny moons of Mars were probably captured intact, it is reasonable to suppose that the creation of Earth's moon was the exception, rather than the rule.

In this chapter we first must determine whether the moon could be closer to the earth today. It might be that the moon is as near the earth as it possibly could be now after orbiting for over 4.5 billion years. In order to be sure that the moon could be closer, we begin by examining why it is moving away.

To recede from the earth, the moon must gain energy from some other body. Otherwise, Newton's law of gravitational force shows that the moon would remain in the same orbit. This additional energy comes from the earth. Through the tides it generates here, the moon actually *compels* the earth to spin it away.

TIDES

Gravitational forces from other astronomical bodies, especially the moon and sun, generate the tides. Since the tides created by the moon are the highest, we will focus on these. The fundamentals of tides are best visualized by temporarily freezing both the earth and moon in space. Imagine that the earth has stopped spinning on its axis, that the moon has stopped orbiting the earth, and that the earth has stopped orbiting the sun. While the earth's spinning and the moon's orbiting are extremely important in understanding both the tides and why the moon is receding, they complicate the basic tidal interaction between the earth and moon. We will put everything back in motion shortly. For the sake of our example, let us imagine that the fixed moon is permanently located over the Atlantic Ocean.

Ocean tides occur because the parts of the earth closest to the moon feel more gravitational pull from it than do the more distant parts. Because the moon is directly over the Atlantic, its waters are closest to the (fixed) moon, and so of all parts of the earth, they feel the greatest gravitational force from it. In response, the Atlantic moves upward toward the moon. This water is lifted higher and higher until the gravitational force from the earth pulls back on it with equal strength and stops its ascent. With the moon and Earth fixed in position, this water would remain at the same high level forever. Water displaced upward like this is called a high tide. This water that rose into the high tide under the moon came from other oceans, which were thereby lowered, creating low tides.

Newton's law of gravity says that the farther apart two objects are the lower the force of gravity they exert on each other. Therefore, the oceans of the earth farther from the moon feel less gravitational force from it than do the ocean and continents directly under it. This might suggest that the lowest tide on Earth should be on exactly the opposite side from the moon. In fact, the lowest tides occur in a ring halfway to the opposite side from the moon where the water was drained away to create high tides elsewhere. On the opposite side of the earth the tide is exactly as high as it is on the side closest to the moon!

The reason there is a second high tide on the opposite side of the earth is because the moon pulls the bulk of the earth away from the oceans on the far side. It is not so much that the oceans rise up there toward the empty sky, but that the earth itself descends away from the oceans and toward the moon. In other words, the solid earth moves away from the oceans on the side opposite the moon by just the same amount that the oceans on the side closest to the moon move away from the earth below them. As a result of the two bulges created by high tide, the earth always has an egg shape. The solid parts of the earth also have tides, but they are much lower than the ocean tides because the rock and dirt do not move as easily as does water.

In our example with the fixed moon and Earth, there would be permanent high tides in the Atlantic and on the opposite side of the globe in the western Pacific (east of Papua New Guinea). At the same time, there would be compensating low tides in a ring between these areas that includes the Gulf of Mexico, the eastern Pacific (near the Galapagos Islands) extending down to Antarctica, the central Indian Ocean all the way up to the Bay of Bengal, and the Arctic Ocean at the north pole.

The Effect of the Earth's Rotation on the Tides

The fixed moon would not rise or set. Rather, it would remain over one side of the earth and would be unseen from the other hemisphere. The ocean water directly below such a stationary moon would be fixed at one height. The high tide on the far side of the earth from the moon would also be fixed in location and height. The moon would *not* move away from the earth under these conditions.

We now put the moon back in motion, while still keeping the earth from spinning on its axis or orbiting the sun. The moon's orbit keeps it over the equatorial regions of the earth. Back in motion, the moon would again be seen from all parts of the earth as it rises and sets. As the moon passes over different parts of the fixed earth, it always pulls hardest on the water closest to it. Therefore the location of the high tide closest to the moon moves in an effort to stay directly under the moon. Since it takes twenty-seven and a third days for the moon to orbit the earth, the ocean water would have plenty of time to adjust itself to the changing position of the moon throughout the month. In other words, if the earth were not rotating, the narrow parts of the egg-shaped earth would always remain aligned with the moon. Under these circumstances, the moon would still not move away from the earth.

We next put the earth back in motion. Today the earth rotates (spins on its axis) about twenty-seven times faster than the moon revolves (orbits) around the earth. The earth's rapid rotation makes the moon rise and set nearly once a day. Just like someone pulling the rug out from under someone else, this rapid rotation pulls the high tide nearest the moon out from underneath it. Specifically, the high tide nearest the moon is pulled ahead of the moon by the earth's motion. The egg is no longer aligned with the moon: The closer high tide, rather than remaining directly under the moon, leads the moon around the earth.

It might seem as if the water in the high tide would rush back under the moon as soon as the earth pulls it away from the line between the centers of the earth and moon. After all, water flows easily. But several factors conspire to prevent this adjustment. First of all, the oceans contain so much water that they have enormous inertia.

Pulled away from the moon by the earth's rotation, the water in the high tide cannot instantly move back under the moon.

Next, as water flows across the ocean bottoms, it rubs against the sand and rock there. This flow generates friction between the water and the material on the ocean floor, which slows the water down. Finally, moving back toward the moon, the high tide inevitably encounters land. If the land is in the form of islands, the water is slowed down as it skirts around these obstacles. If the land is a continent, the tide rises up on the beach and pushes on the land. When the land does not move out of the way, the tide eventually subsides, never having attained a position directly under the moon.

The Effect of Ocean Tides on the Earth's Rotation

In attempting to stay under the moon, the high ocean tide flows east to west, pushing against the ocean bottoms and running head-on into the east coasts of the continents. Since the earth rotates from west to east, the tidal ocean water is exerting resistance on the solid earth, slowing the earth's rotation rate.

We are now in a position to understand why the moon is moving away. The answer centers on the tidal bulges (high tides) created by the moon on the rotating earth and the fact that the bulge nearest the moon is not directly under it.

The Effect of Ocean Tides on the Moon

The gravitational pull on the moon by the egg-shaped earth can be divided into three parts: Most of the earth's pull comes from its central mass, while the two bulges pull on the moon separately. If the narrow parts of the egg-shaped earth were always aligned with the moon, the two bulges and the main body of the earth would all pull the moon toward the earth's center. In that case Newton's law of gravitation indicates that the orbit of the moon would be an ellipse and that it would remain in that orbit forever. But the tidal bulge nearest the moon leads that body around in its orbit; the narrow part of the egg points ahead of the moon.

The two tidal bulges are masses of water located at different distances from the moon. Leading the moon in its orbit, the closer tidal bulge pulls harder on the moon than does the tidal bulge behind the earth. Clearly, these two tidal bulges pull on the moon in different directions, neither of which is toward the center of the earth.

The tidal bulge closer to the moon continuously pulls the moon forward in its orbit and speeds it up. The more distant tidal bulge trails the moon and therefore pulls back on it, slowing it down. However, since this latter bulge is farther away from the moon, it has less effect than the closer bulge. The net result is that the moon feels a force from the ocean tides pulling it forward and speeding it up. Speeding up has the effect of making the moon spiral away from the earth.

The Effects of the Sun on the Tides

Finally we put the earth and moon back in orbit around the sun. When the moon and sun are aligned, their forces act together and thereby create extremely high tides on Earth, the so-called spring tides mentioned in chapter 1. Since the sun and moon each create pairs of high tides on opposite sides of the earth, spring tides occur when the moon is on the opposite side of the earth from the sun, as well as when it is on the same side.

When the moon and sun are at right angles from each other as seen from the earth, their gravitational forces oppose each other here, and the resulting tides they generate partially cancel each other. This creates the lowest tide cycles of the month, called neap tides.

We can summarize the total gravitational interaction between the earth and moon this way: The moon draws the earth into an egg shape by creating two high ocean tides. The earth drags these tides around with it. In response, the tides push against the solid earth and thereby slow the earth's rotation. At the same time, the gravitational force from the closer high tide pulls the moon ahead in its orbit. This speeds up the moon and sends it spiraling outward. The tides from the sun and moon sometimes reinforce each other, creating especially high tides, and sometimes oppose each other, creating especially low tides.

WHERE THE CLOSER MOON MIGHT BE LOCATED

Knowing why the moon recedes from the earth, we can now explore the question of whether it could be closer today. Certainly if the moon originally orbited closer to the earth, it would take longer to spiral outward to its present distance and would therefore be closer today. But we cannot start the moon in an orbit arbitrarily close to the earth. There is a minimum distance between the earth and the moon that must be maintained. This distance is called the Roche limit, after M. Edouard Roche, a French mathematician who first calculated it in 1847. If our moon started life at the Roche limit, it would be hard (although not impossible) to justify that a moon of the same mass could be closer today. After discussing the Roche limit, we will examine our present moon's orbital history to see how far away it was initially and whether it somehow could have receded more slowly.

Minimum Distance from the Earth to the Moon— The Roche Limit

The Roche limit exists because the earth creates tides on the moon in the same way that the moon creates tides on

Earth. To avoid confusion, we will hereafter refer to the tides on the moon as solid tides. As on Earth, closer parts of the moon feel more gravitational pull from the earth than do more distant parts. However, because the moon's surface is solid, it flexes less in response to the earth's tidal force than our oceans do in response to the moon; the moon's solid tides are smaller than the ocean tides on Earth.

In the past, when the moon was closer to the earth, the earth's tidal effect on it was greater, creating higher solid tides on its surface. If the moon had ever been as close as the Roche limit, the earth's gravity would have been strong enough to create such high solid tides on it that its surface would have been literally lifted into space. The newly exposed interior of the moon would have become its surface, and it too would have been lifted away by the earth's gravitational pull. At the Roche limit the moon would be peeled apart like an onion until there was nothing left of it but a ring of rubble orbiting the earth. We can look at the Roche limit this way: It is the distance from the earth at which the solid tides on the moon would be so high that the moon's own gravity would lack the strength necessary to hold them down. If the debris from which the moon originally formed had begun orbiting the earth inside the Roche limit, that material would never have coalesced to form the moon in the first place, and Earth would have had a permanent ring.

Every planet has its own Roche limit. Saturn, Jupiter, Uranus, and Neptune all have rings that are inside their respective Roche limits. These rings can never become moons. For example, the rings of Saturn range from 50,000 to 82,000 miles from that planet's center. They are all inside Saturn's Roche limit, which is located 90,000 miles from the planet's center. Also, all the moons of these planets (except for a few that are irregular in shape and held together by atomic forces rather than by their own gravity) are outside the Roche limits.

The earth's Roche limit is approximately 11,000 miles from its center, or 7,300 miles above the earth's surface. The moon is presently 230,000 miles away. This is twenty times farther away than the Roche limit. The moon began its orbit somewhere between the Roche limit and its present location, but where?

Tectonic Plates and Drifting Continents

We have seen that the rate at which the moon is spiraling away from the earth is linked to the rate that the earth is spinning downward. The earth's tidal bulges connect the two phenomena. The water in the tidal bulges slows the earth by rubbing against the ocean bottoms and pushing on the continents. To determine the moon's initial location, it is necessary to know the exact contours of the ocean bottoms and to locate where the continents have been throughout the earth's life. That would tell us how much the ocean bottoms and continents have interfered with the oceans' tidal flow and how much the earth's rotation has changed because of it.

We measure this change in rotation as a loss of angular momentum by the earth. Angular momentum signifies the difficulty involved in changing an object's rate of rotation or revolution. The heavier or larger a rotating object is, the greater its angular momentum. Angular momentum is always conserved; if the earth loses angular momentum due to tidal friction, the moon must gain the same amount. By Newton's law of forces, the increase in the moon's angular momentum leads to an increase in the moon's orbital speed around the earth. Such an increase in speed leads to an increase in distance from the earth to the moon—the moon spirals away. When we use the mathematics of angular-momentum conservation, we can determine how far outward the moon has moved and, therefore, how close it was when it first formed.

Unfortunately, we cannot tell directly how much the continents have interfered with ocean tides throughout all of the earth's history. The continents are moving in both longitude and latitude, as well as relative to one another. This has been going on for at least the past 2.8 billion years. The continents drift because they are a series of rock icebergs resting on and moving over the earth's mantle, which is composed of rock that is easily deformed. The continental "rockbergs" are called tectonic plates. The surface of the earth is completely covered with these plates; they form its crust.

The earth's interior is composed partially of molten rock, called magma. Magma escapes to the earth's surface (or ocean bottoms) in the form of lava. This magma is, among other things, the source of volcanoes.

Seeping upward between tectonic plates, the magma pushes the adjacent pieces of the earth's crust aside. As a result of the magma acting on them, the tectonic plates are forced to migrate. Sometimes plates are moving apart, like those containing Europe and Africa, which are now drifting away from the Americas. Some plates drift toward each other, as the Americas and Asia are doing. Sometimes plates collide with each other. The Indian subcontinent is presently ramming into the underbelly of Asia, thereby creating the Himalayas. Sometimes plates scrape against each other, as is happening on the west coast of North America along the San Andreas fault. Tectonic plates are always coalescing, breaking apart, and reforming with new shapes. As a result, the continents are continually changing shape and location.

Geologists have traced the paths of the continents as far back in time as about 1.5 billion years, but they lack the details of locations and speeds. What happened before then is still shrouded in the haze of time.

Our Present Moon's Orbital History

Since we lack precise information about continental motion throughout the earth's entire history, we cannot know for sure how much the ocean bottoms and continents have affected the flow of the tides and thereby slowed down the earth. Were there periods when water flow near the earth's equator was free from interference by continents? At such times the earth's tidal bulges would have moved more easily than they do now. They would have slowed the earth down less and sped the moon up less than they do now when continents are in the water's path. Without the full history of continental drift, we cannot be sure how much the earth has slowed down, at what distance from the earth the moon first started orbiting, or when the moon first formed.

And yet, there is still a way of deriving the moon's original distance from Earth. Instead of working backward from the present, we can work forward from the creation of the moon. We can derive reasonable estimates of the distance above the earth at which the moon coalesced by using sophisticated computer models of the earth. Such models were originally developed for studying how shock waves from nuclear explosions would travel through the earth. In order to obtain realistic results about the earth's response to nuclear tests or nuclear war, these models contain many details of our planet's interior, surface structures, and gravitational field. Using these programs, geologists can simulate the impact on the earth of the Mars-sized planetesimal that formed the moon. Such simulations show matter being splashed off the earth's surface and projected into orbit around the planet.

Simulations of planetesimals with a variety of different masses, speeds, and directions were tested on the model earth in the computer. For example, when material was splashed off the earth into an orbit that was initially inside the Roche limit, no moon formed. The debris remained in a ring. However, a moon did form when a collision put material in orbit outside the Roche limit. From these computer experiments, we can derive a plausible scenario for the moon's formation.

When we program a collision between a Mars-sized planetesimal and the young earth on the computer, we can see debris being thrown into an orbit around the earth one and a half times farther away than the Roche limit. Lacking a better indicator, we assume that the moon actually began orbiting the earth at this distance. Therefore, since its formation the moon has moved outward at least thirteen times its original distance from the earth! Bearing this information about the birth of the moon in mind, we are now ready to move it closer for this chapter's new world.

How the Moon Could Be Closer Today

The moon could be closer today for any of three reasons. First, it could have formed later in the history of the solar system than it actually did. In that case, it would have had less time to move away from the earth. Second, we have just seen that the moon could have begun orbiting closer to the earth, perhaps just outside the Roche limit. That way the moon would have had farther to recede, which would take more time than it actually took. Third, the rate at which the moon moved away from the earth could have been slower. This would have occurred if the continents had interfered less with the tidal water flow than they actually did. The tidal bulge nearest the moon could then remain more directly underneath it. By not leading the moon as much, this tidal water would then have pulled the moon less ahead of itself, slowing the rate at which the moon spiraled outward.

For the scenario of this chapter, I assume that all three

conditions occurred: That the closer moon formed later than our moon, that it formed nearer the Roche limit than our moon, and that the earth's continents stayed near the poles through more of the earth's history than they actually did. This assures that the closer moon's recession rate (the rate it moves away from the earth) would have been slower than the recession rate for our moon.

We will call the earth with a closer moon Lunholm. For convenience in visualizing the effects of Lunholm's closer moon, we will assume that Lunholm's continents are now located in the same places as the continents of Earth and that Lunholm now has a twenty-four-hour day. From the discussion in chapter 1, it might seem that the length of day on Lunholm must now be different from twenty-four hours. After all, both Earth and Lunholm formed with six-hour days. To slow down to a twenty-four-hour day, Lunholm would have to lose as much angular momentum as the earth.

If the moon-forming planetesimal hit Lunholm at a different angle than the angle at which the planetesimal hit the earth, it would be possible to have a twenty-four-hour day on Lunholm today. By changing the direction of the collision, we would change the amount of angular momentum transferred between the planetesimal and the planet. This, in turn, would change the rotation rate of Lunholm. Depending on whether the impact was head-on or merely glancing, the rotation rate on Lunholm could have been very different from the post-impact earth. Therefore, we assume that Lunholm was slowed down greatly by both the impact and by subsequent tidal interaction with its closer moon.

Lunholm's moon, while otherwise identical to ours, is located one quarter of the distance from the earth to the moon today. The closer moon's diameter would appear four times larger in Lunholm's sky than the moon appears to us.

The Closer Moon's Orbit

The time it takes the moon to orbit the earth or Lunholm can be measured in two completely different ways. First, the orbit can be taken from full moon to full moon, called the moon's synodic orbital period. *Synodic* means with respect to the sun. The full moon occurs when the moon is exactly on the opposite side of the earth from the sun. The phases of the moon are discussed in detail shortly. Our moon's synodic orbital period today is twenty-nine and a half days long.

Alternatively, the moon's orbit can be measured from when it passes a certain (arbitrarily chosen) star in the sky until the next time it passes that same star. This is called the moon's sidereal orbital period. *Sidereal* means with respect to the stars. Our moon's sidereal orbital period today is twenty-seven and a third days long.

If the earth and sun were fixed in space, the synodic and sidereal orbital periods would be the same. They are different because the earth orbits the sun: As the earth moves, the sun appears to move among the background stars. Suppose that the orbital periods of the moon are measured starting when it is full. At that time, the stars behind it are carefully noted. One sidereal orbital period of the moon later (twenty-seven and a third days), the moon is back among the same background stars, but since the earth has moved around the sun, the moon is not yet full again. The moon has to orbit a little over two days more before it becomes full. This completes its synodic orbital period (when the moon gets directly behind the earth from the sun again). We will use both types of orbital periods throughout this chapter.

Kepler's third law determines the moon's sidereal orbital period around the earth. This law asserts that the closer the moon is to the earth, the shorter its sidereal orbital period. The same relationship between distance and orbital period applies to the sun and planets. Namely, planets closer to the sun have shorter sidereal orbital periods around it than do the more distant planets. (Interestingly, in both cases the mass of the orbiting body has no effect on the period of the orbit.)

Kepler's third law shows that the closer moon would orbit Lunholm more rapidly than our moon orbits the earth. Instead of orbiting in twenty-seven and a third days, Lunholm's moon would have a sidereal orbit of three and a half days. The associated synodic orbital period (cycle of phases) would be nearly four days long. Most noticeable of the closer moon's visual effects would be the creation of more frequent eclipses on Lunholm.

Eclipses

Eclipses arise because planets and moons create cone-shaped shadows of darkness on their night sides. The shadow cone's base is the planet or moon. The cone tapers to a point in space aiming away from the sun. The earth or Lunholm passes through the moon's shadow cone when the moon is exactly between either of them and the sun. Our moon at its present distance is so far away that only the very tip of its shadow cone ever touches the earth. People standing on the earth as the moon's shadow races by would see the sun disappear behind the moon for no more than seven minutes. This is a solar eclipse. People who are not within a region a few hundred miles wide where the moon's shadow cone touches the earth would miss the solar eclipse.

Conversely, the moon passes through the earth's or Lunholm's shadow cone when it is exactly on the opposite side of the planet from the sun. This creates a lunar eclipse. Because the earth's shadow cone is so large, it usually engulfs the entire moon during a lunar eclipse. In principle, everyone on the side of the earth facing the moon at that time can see the eclipse.

It might seem that solar eclipses should occur every time the moon passes between the earth and sun (new moon) and that lunar eclipses should occur every time the moon is full. This does not happen because the moon orbits the earth in a slightly different plane than the earth orbits the sun. The earth orbits the sun in a plane called the ecliptic. The plane of the moon's orbit around the earth or Lunholm is tilted by five degrees in relation to the plane of the ecliptic. This inclination is enough to prevent the moon's shadow cone from hitting the earth by passing above or below it during most new moons. Likewise, the moon usually avoids entering the earth's shadow cone by passing above or below the cone when it is full.

When the new or full moon occurs with the moon close to the ecliptic, an eclipse occurs. Today a maximum of seven and a minimum of two eclipses occur per year, with no more than five solar eclipses or three lunar eclipses. Even though Lunholm's closer moon would still orbit in a plane tilted from Lunholm's ecliptic, that moon would be close enough to encounter Lunholm's shadow during virtually every synodic orbit. Similarly, the moon's shadow cone would strike Lunholm at every new moon. Recall that a synodic period on Lunholm, coinciding with a cycle of lunar phases, would be only four days long. Therefore Lunholm would experience nearly 180 eclipses each year (90 of each kind), rather than the 2 to 7 eclipses we have now on Earth.

Protection of Lunholm By the Moon

The closer moon would bring about other changes as well. Because the moon would be larger in Lunholm's sky it would provide the planet with greater protection from incoming planetesimals than our moon provides the earth. Like a mammoth shield orbiting Lunholm, the closer moon would be in the path of some of the incoming space debris that would otherwise hit the planet. Such protection would be invaluable throughout the life of any planet, especially one that carries precious living cargo.

Although the vast majority of the planetesimals and smaller pieces of space debris in the solar system struck planets and moons billions of years ago, some still remain. Among the millions of pieces of debris that the earth's gravity has pulled down just within the past twenty thousand years, at least one was one hundred feet across. This latter rock struck the earth and exploded with the energy of 4 million pounds of TNT. Its impact formed a crater near present-day Flagstaff, Arizona, that is nearly a mile wide and six hundred feet deep. Although these impacts are too weak to create another moon in the way we discussed in the first chapter, they can still cause enormous damage.

Our moon has provided the earth with a modicum of protection. We see evidence of this in the heavy cratering that occurred on the far side of the moon. Since it never faces the earth, that side of the moon has taken the brunt of untold thousands of collisions from space debris that would otherwise have struck us. Many other large pieces of space debris missed our moon, however. A few miles across, each of these struck the earth with more energy than the combined explosive power of all the nuclear weapons ever built. The debris from at least nineteen major impacts during the past 500 million years has been identified. If even one of these events had been avoided, both the time line and direction that evolution took would be profoundly different than they were on our earth.

Compare, for example, the impact that caused all the dinosaurs to perish 65 million years ago with what might have occurred had that object struck the moon instead. The piece of space debris that caused the so-called mass extinction of the dinosaurs and many other species of animals and plants is believed to have been less than fifty miles across. Its impact caused earthquakes and tidal waves incomparably larger than anything in human experience. It also blasted thousands of cubic miles of dust into the air. Blocking sunlight, this dust cooled off the earth for years or perhaps even decades. The darkness created by the dust killed vegetation, froze many animals to death, and generally disrupted the food chain. Many species were wiped out as their habitats became unlivable.

The evidence of this event comes from chemicals removed from the piece of space debris by its impact and spread over the earth by the subsequent explosion. The object that struck the earth contained several chemicals that are rare in the earth's surface, most notably iridium. Very high concentrations of iridium have been discovered all around the world in the rock layer that was the earth's surface 65 million years ago. At Chicxulub, on the Yucatan Peninsula in Mexico, geologists have also found debris from an impact crater that is believed to be associated with this extinction.

As a result of the mighty dinosaurs disappearing, the tiny mammals then alive became larger and more diversified. Had this impact occurred on the moon instead of the earth (or Lunholm), the subsequent mass extinction of life would not have occurred. In that case, the evolution of humans from those early mammals would have been delayed or even prevented as dinosaurs continued to flourish and fill the biological niches that mammals filled on our earth after the dinosaurs were gone.

Ocean Tides

The closer moon's greatest effect on Lunholm would be to generate exceptionally high ocean tides. The heights of ocean tides are extremely sensitive to the distance between the earth and the moon. Ignoring all effects other than gravity, if the moon were half as far away from the earth as it is, the tides here would be eight times higher than they actually are. At one quarter of its present distance, the moon would create tides on Lunholm sixty-four times higher than they are today. Because Lunholm's moon would orbit that planet faster than the moon orbits the earth, the tides on Lunholm would be even higher still.

To understand this increase in tidal heights, note that the moon is lumbering around the earth very slowly (once every twenty-seven and a third days) compared with how rapidly the earth rotates (once every twenty-four hours). Therefore the moon spends very little time above any one part of the earth each day. As the high tide on Earth rises up toward the moon, the moon is rapidly slipping behind it. Therefore, tides on Earth don't have time to build up as high as they might if the moon remained longer above each region of the ocean.

Since Lunholm's moon orbits once every three and a half days, Lunholm would not be spinning as fast relative to its moon as the earth spins relative to our moon. Each time the closer moon orbits Lunholm, that planet has rotated only three and a half times below it. The closer moon would remain above each part of Lunholm longer than our moon does above each part of the earth. Because of this, the tides on Lunholm would have more time to rise up toward the moon before they are pulled away from it by the planet's rotation. As a result, the tides on Lunholm would rise more than sixty-four times the height due to the closer moon's increased gravitational force on the planet. Tides seventy or eighty times higher than those on Earth would be the norm on Lunholm. Such tides would profoundly affect the shores of the planet.

From a human standpoint, the most important effect of such monstrous tides would be on the inhabitability of Lunholm's coasts. Most coastal cities on the earth are built comfortably above the reach of the highest tides. Like New Orleans, low-lying communities fend off the ocean with dikes or levies that surround them and rise above the high-water line. Except in rare cases when storms whip waves over protective barriers, coasts present accessible, convenient, attractive, and safe places for many of Earth's largest cities. Also, ocean levels usually rise or fall so slowly around the earth that people can play on most beaches throughout a complete cycle of tides. Besides sunburn, the worst problem we usually face during a day at the beach is having to move our blankets a few yards inland as the water rises.

We can gain some insight into the tidal dynamics on Lunholm's shores by examining the relatively rare beaches on Earth that have extremely gentle slopes compared to the height that the tide rises. While they are found all around the world, perhaps the best known of these shores surrounds Mont-Saint-Michel on France's northern coast. The restless ocean can move the shoreline a mile or more in just a few minutes in such places. Such beaches are notorious for quickly disappearing as the tide rises, taking with them any hapless beachcomber who might inadvertently wander out too far. On Lunholm, tides would typically be several hundred feet high. Because they are so high, these tides would have the same effect of covering the beach along virtually every shore of that world, no matter how steep. Lunholm's oceans would rush miles, often tens or even hundreds of miles, inland with each rising tide.

The oceans would scour millions of square miles of what on Earth is coastal lowland, making this area uninhabitable by humans. Some tides would rise gradually; others would be more aggressive. Pushed by the waters behind them or channeled by the ocean bottoms, some rising tides would create tidal waves several hundred feet high. This water would roar inland, its unassailable wall crushing, dragging, and destroying everything in its path.

How high the tide reaches and how forcefully it surges against each shore would depend on the distances from each shoreline to the end of the continental shelf, as well as on the shapes of ocean bottoms, straits, and coastlines. The continental shelf is the underwater edge of each continent, beyond which the ocean bottom falls off precipitously. Generally, the farther the shore is from the continental shelf, the higher and more powerful are the tides. The highest tides on Earth are found in the Bay of Fundy, where they have exceeded fifty feet in a single cycle of tides. The lowest tides are found, among other places, in the Mediterranean Sea, the Gulf of Mexico, and in the mid-Pacific, where the variations throughout a cycle are often less than two feet.

By moving vast quantities of debris, the ocean tides of Lunholm would move and reshape shorelines much more rapidly than do the tides on Earth. Sometimes the deposits would be moved inshore, increasing the height of beaches; sometimes they would be moved offshore, lowering the shore land. Taking place over only a matter of years, these changes would move shorelines by miles or more each century.

Rocky coastlines like those of Maine, while more resistant to change, would decay faster than they do on Earth. Rocks would break up under the relentless pounding of the tidal water. The white-chalk cliffs of Dover, England, would erode much more rapidly than they do today, greatly widening the English Channel. Even if the coasts of Lunholm were exactly identical to those of Earth today, in a few years they would be unrecognizable to us.

River Tides and Bores

Rivers, too, would be greatly affected by tides on Lunholm. On Earth many large rivers that flow sedately to the sea would be raised out of their banks by the upflowing tidal water on Lunholm. Vast floodplains, each covering thousands of square miles along the riverbanks, would fill and drain with each ebb and flow of the closer moon's enormous tidal pull. This land adjoining rivers is among the most fertile on Earth. On Lunholm, the salt water deposited on the land by the oceans would severely limit the types and quantity of plant growth that could live in these regions.

Like Earth, Lunholm may have tidal bores—river tidal waves. Whether or not a bore occurs depends on the shape of the riverbed, the shape of the bay into which the river empties, and the distance between the mouth of the river and the edge of the continental shelf. If the riverbanks there are high enough to support them, the bores on Lunholm could be hundreds of feet tall. On Earth, tidal bores can be up to twenty feet high but are usually much lower. Today bores occur on the Saint John River between Maine and Nova Scotia, on the Severn River between Wales and England, and on the Seine, Orne, and Gironde rivers in France, among others.

While Lunholm's oceans would erode the shores, its tidal bores would dramatically erode riverbeds and land adjacent to the rivers. The lands astride Lunholm's tidal rivers, therefore, would be completely unusable for farming and living.

Icebergs

Tides would erode more than just rock and sand along the shores on Lunholm. As on Earth, Lunholm's polar seas would contain floating ice and abut mighty glaciers—layers of ice and snow that build up on land from centuries of snowfall. As a glacier thickens, its lower levels are compressed by the weight of the snow above until they turn into ice. Glaciers flow downhill. While some glaciers, including many in the Alps, are located in mountain passes far from the sea, others descend right into an ocean, such as some of those found in Alaska and Greenland. There are also ice shelves extending from the land into the water, most notably the Ross ice shelf in Antarctica.

As the huge ocean tides and river bores on Lunholm surge against the brittle ice faces of the glaciers and shelves, titanic pieces of ice would snap off and become icebergs littering Lunholm's seas. Traveling with ocean currents and winds, many of these frigid rafts would float toward the equator. The larger ones would be broken into smaller pieces by the tempestuous oceans. Floating ice would frequently issue into the Pacific through the Bering Strait, between Alaska and Siberia, and from Antarctica. The narrower Atlantic would be full of icebergs from the seas around Greenland in the north and from Antarctica in the south.

Icebergs would occasionally be washed ashore. Carried by tides, some of them would end up high on beaches at or beyond high-tide lines. This is reminiscent of the beaching of some boats miles inland during hurricanes on Earth. Since these icebergs would contain mostly fresh water, they could be mined for drinking water in times of drought.

Solid Tides

While we never see it happen, the solid parts of the earth today experience tidal motion due to our moon's gravitational force. The land on Earth rises and settles throughout each cycle of lunar phases, although the greatest tidal displacement of land reaches only eight inches. On Lunholm, however, some landmasses would move as much as ten feet at high tide under the closer moon's influence. Such displacements of the continents would create enormous stresses in the rocks of Lunholm's crust. These stresses would lead to frequent, powerful lunholmquakes as the land periodically readjusts itself to the changing forces acting on it. By comparison, geologists believe that the tidal motion generated by our moon is too weak to cause earthquakes directly.

The tidal stresses in Lunholm's crust would be most acute at the boundaries between tectonic plates, where the rock is already under great pressure. The frequent displacements due to the tides would serve to create greater tectonic activity by making it easier for the plates to disturb one another.

Those tectonic boundaries along Lunholm's coastlines would face another potent danger in the form of the greater tidal water flow. Tidal waves would continually wash over such quake zones as the San Andreas Fault in California. The persistent ebb and flow of the oceans would wear down the land above these faults more rapidly than occurs now, thereby exposing the unstable rock below. Combined with a greater weight of water bearing down on the faults and the tidal motion in the rock itself, this exposure would cause much more strain than exists there today. As a result, quake activity along fault lines would be far more common than on Earth. In fact, many coasts on Lunholm would probably have constant seismic activity, perhaps accompanied by continuous volcanic activity as magma leaks out between the plates.

LIFE ON LUNHOLM

Accelerated Evolution

The turmoil on Lunholm created by the closer moon's gravitational force would set in motion events that could speed up the initial spread of life in the oceans. We recall from chapter 1 that lower tides on moonless Solon would bring fewer minerals into the ocean and spread them more slowly than occurred on Earth, delaying the development of early life forms.

Just the opposite would happen on Lunholm. Roaring miles overland, the enormous tides would retreat back to Lunholm's oceans carrying vast amounts of minerals. The ready availability of these chemicals in the oceans would hasten the initial stages of evolution by providing more building blocks of life. The first life forms would also have a richer chemical environment in which to replicate than they had on our earth or on Solon.

We also saw in chapter 1 that tidal water flow, the uncovering of thousands of square miles of beaches at low tide, and vast quantities of chemicals necessary to duplicate life were all major ingredients necessary for the reproduction of the earliest life forms. The closer moon that orbits Lunholm would provide more of these things than did our moon. This moon would generate an explosive burst of reproductive activity on Lunholm exceeding the rate that occurred on our earth. (Keep in mind that the earth's early tides were huge by today's standards because the moon was much closer then than it is now. However, the tides would always be higher on Lunholm than on Earth since Lunholm's moon started orbiting closer to that planet than our moon started orbiting Earth.)

The more abundant bacterial and algal life in the early oceans of Lunholm would convert carbon dioxide into oxygen more rapidly than occurred on our earth. This increased oxygen content in Lunholm's early oceans would then allow for the evolution of aquatic oxygen-consuming animals. The plentiful early oxygen would also hasten the conversion of Lunholm's carbon dioxide–dominated atmosphere to the one we breathe now, containing nitrogen and oxygen, as discussed in chapter 1. Lunholm's surface would be primed for the development of animal life long before the earth was. However, an evolutionary bottleneck would exist that would make it difficult for animal life to escape onto land from Lunholm's teeming and violent oceans.

Decelerated Evolution

Land animals are descendants of aquatic life. On Earth, the transition to life on land came when amphibians crawled or were deposited on beaches by tides nearly 400 million years ago. These creatures primarily breathed oxygen dissolved in water through their gills. They had limited capacity to breathe the oxygen-rich air on the earth's surface. As discussed in chapter 1, oxygen in high concentrations is toxic, making the evolution of lungs that could limit the amount of oxygen entering an animal's bloodstream an important development. Fortunately for our ancestors on Earth, the ocean tides 400 million years ago had decreased to little more than they are today. Therefore our ancestors emerged from the oceans during relatively calm high tides, stayed on land a few minutes, and returned to the ocean before the water receded too far for their flipper legs to carry them.

The tides on Lunholm would make the transition to the land considerably more difficult. Whereas most tides on the earth of 400 million years ago were relatively sedate, most tides on Lunholm at that time would still be powerful, towering flows. In many places the amphibians near the leading edge of a tide would be in mortal danger of being scraped along the ocean bottom, smashed against rocks on the shore, or tossed so far inland that they would be unable to return to the water before dying from breathing too much oxygen.

As on Earth, there would be great variety in Lunholm's shorelines. Just as there are places on the earth where the tides are much lower than they should be (due to the shape and depth of the ocean), there would be places on Lunholm where the tides would be tempered sufficiently by the topography of the land for amphibians to venture safely ashore. Because there would be many fewer such places on Lunholm than there are on Earth, the emergence of life on land would experience a bottleneck. The appropriate amphibians would have to be in these calmer waters for the transition to land to occur. With the successful evolution onto land, the difficulties of shore life would only be beginning.

Shore Life

On Earth there is a breathtaking variety of shore life inside the intertidal part of every beach—the region covered at high tide and uncovered at low tide. Many aquatic species also live in the shallow tidal water, either moving with the water flow or fixed to the bottom of the tidal land. The violent tidal waters on Lunholm would require shore life to develop even better adaptations than were necessary for shore creatures on Earth.

The animals living on Lunholm's intertidal ocean floors would have to be very firmly attached to rocks. Otherwise, they would be ripped from their moorings by the pounding they would receive from waves and waterborne debris. Rocks insecurely embedded in the shore would often be moved and the animals living on them would be battered as their homes tumbled against other stones. The thickness and shapes of shells would make all the difference in such a violent world. For example, creatures such as barnacles that are relatively flat, well secured, and ruggedly constructed would probably do well on Lunholm's shores. Mussels, with their thinner, more easily broken shells, would be less successful; moreover, they are connected to rock along their thin edge so that they stick upright. They could be broken off or crushed relatively easily when struck by tidal debris.

Tidal creatures such as clams that are buried in the mud would have to burrow even deeper down to prevent themselves from being dragged away or crushed as the tides scour the shore. Unfortunately, fish that live near the ocean's surface would be particularly unsuited for coastal life on Lunholm. Thrown ashore, bashed into rocks, grated

against the ocean bottoms by the tidal currents, these fish would stand little chance of surviving there. Perhaps armor-plated fish, able to roll up in a ball as they are thrown about by the waves, would evolve to help fill the challenging biological niche of the intertidal zone. Generally, that is one biological niche on Lunholm that would be much less habitable than its counterpart on Earth.

Fish like salmon that migrate from streams to the ocean and back would face even more daunting challenges than they do on Earth. After being born in freshwater streams, these fish swim downstream toward the ocean. Once adapted to salt water, they have little difficulty fighting tidal flow in order to enter the ocean. When they are ready to spawn, they are able to locate their home river and swim back up it against the current.

On Lunholm such fish would face greater tribulations in both directions. On the way downstream they would encounter powerful tides and tidal bores throwing them out of the water or against the bottom of the rivers. To successfully get to sea they would have to time their arrival to coincide with the calmest period of high tide. Even then they would have to swim as hard as they could to get to deep water before the receding water scraped them against the bottom of the ocean's tidal region. These fish would have great trouble finding their rivers. When they come home to spawn, the tidal motion of the water would rapidly disperse the distinctive chemicals that each river emits in the oceans, making it harder for these fish to locate the homes. One alternative for such fish on Lunholm is ascend any river they encounter to spawn, rather than j the one from which they came.

Tectonic-Plate Boundaries and Life

As discussed earlier, Lunholm's closer moon would greater quake activity, especially along the boundari

tectonic plates. While life would probably be absent from regions of constant lava flow, areas of Lunholm's surface near nonvolcanic plate boundaries would provide particularly challenging habitats. Animals living in these areas would need special adaptations to help them cope with rapid movement of land, impact from rocks, inundation by water, burial from landslides, and sudden formation of crevices.

Many adaptations, some already present in various animals on Earth, would be useful in quake-prone places on Lunholm. Hard shells would be useful protection against tumbling rocks. Long arms or tentacles, able to cling to vertical rock faces, would help when falling into a crevice. Animals able to hold their breath long enough to dig themselves to the surface after being buried would have an evolutionary advantage. Ultrafast reflexes would also be especially useful in running to safety, dodging boulders, and climbing out of crumbling crevices. Exceptional sensitivity to vibration would serve as an early warning system.

While dangerous, tectonic boundary regions thousands of miles long would present numerous opportunities for the evolution of quakeproof animals on Lunholm. Who knows, perhaps their need to respond quickly might even enable one species among them to evolve the complex brain, opposable thumbs, and other features that would make them analogous to humans on Earth.

Nocturnal Animals

Lunholm's closer moon would shine sixteen times more light onto that planet's surface than our moon shines onto the earth. When the closer moon is full, it would shine brightly enough to keep the entire sky blue—not as bright as the sun makes it, of course, but blue and bright enough so that most stars visible today would be washed out by the light.

The dark of night would be a less effective ally to other-wise vulnerable prey on Lunholm. Also, the greater nightly illumination would require nocturnal animals to have better camouflage. This could be achieved in several ways. Animal coats might more closely match the color and texture of the land on which the animals live. However, this adaptation would limit the range of many animals who would then be more conspicuous on different terrain. Also, the development of a more specialized coat would endanger a species when its habitat changes as a result of fire, drought, or new vegetation.

Alternatively, vulnerable animals might develop skins that could rapidly change color depending on the surroundings. Like chameleons, various fish, squid, octopi, and other land-animal species today, they would then be able to change coloration quickly to adapt to new territories or different lighting. Perhaps Lunholm would be inhabited with more dynamically changing nocturnal animals, whose coats, skins, smells, and noises would vary in order to better help them survive.

Greater Strength and Faster Brains

With its colossal ocean tides and more frequent quakes, Lunholm would be a much less inviting place to live than our earth. Fortunately, nothing obvious in these differences would completely prevent complex life from evolving on Lunholm. The distinctions between Lunholm and Earth are matters of degree, rather than of type. Lunholm's less stable surface might force evolution to equip many animal species with faster reflexes and better instincts for dealing with such calamities as quakes than animals have on Earth. All large creatures on Lunholm would benefit from greater speed, stronger muscles, keener sight, and better coordination than are required for animals on Earth.

Powerful quakes on Lunholm would threaten lumbering beasts unable to get out of danger quickly. For example, large, slow-moving dinosaurs would be at a distinct evolutionary disadvantage in this respect. It is likely that on Lunholm ancestors of humans or their counterparts would be among the species with highly developed responses to quakes. The need for quicker reflexes than our ancestors needed on Earth suggests that our counterparts on Lunholm would develop different brain structures to support this need. As a result of responding more quickly to physical challenges, their higher brain functions, such as logical reasoning, would also have to take place more rapidly.

This connection in speed between the lower-level (reflex) and higher-level brain functions is not arbitrary. Consider what would happen in the encounter of two human strangers on Lunholm. Like us and virtually all other creatures on Earth, they would initially have a "fight-or-flight" response to each other. This instinct protects animals who find themselves in dangerous situations by driving them either to fight their supposed enemies or flee. If the stronger and more agile people on Lunholm did not possess the ability to interpret the behaviors of strangers rapidly, their fast reflexes would lead either to dangerous confrontations or rapid departures.

More combat would lead to more injury and more premature death. Such interactions would decimate species. However, if people on Lunholm developed the propensity to avoid one another instead of to fight, they would be less likely to work together and develop stable communities. In either case, they would end up being less sociable and perhaps less socializable than humans are on Earth. The development of faster brain functions, then, is a plausible evolutionary step for creatures with greater speed and strength than those that evolved here on Earth.

Discovering the Americas

Christopher Columbus was making waves in the fifteenth century with his proposal that the Orient could be reached by sailing westward over the ocean. By the end of that century, he had (re)discovered the Americas and returned home safely, bringing news of both the new land and the peoples living there. Had Columbus lived during the equivalent era of expanding global markets on Lunholm, he would never have dreamed of sailing westward. Because of treacherous tides, ships of any kind on Lunholm would be confined to some of the smaller lakes of that world.

Humans on Lunholm would be restricted to the greater European/Asian/African landmass right through the beginning of the twentieth century. The "globe" would be mostly covered with oceans. Missing from the "known world" of the early twentieth century on Lunholm would be the Americas, Australia, and Antarctica. As steel-hulled ships improved, ventures further afield would be made. Plausibly the best such ships on Lunholm might discover the Japanese islands, the Philippines, and Madagascar. If the tides and waves did not keep mariners off most of Lunholm's oceans, the thousands of icebergs infesting the oceans probably would. Even with the advent of sonar and radar to help keep track of the bergs, the high seas would be exceptionally dangerous.

Many of the animal and planet species found in the pristine Americas of Lunholm by the first human explorers would be related to species known on their home continents. While the American species would often have some different characteristics, they would have the same genetic origins as their Eurasian and African counterparts. However, the likelihood that even one species on the Americas would evolve the same genetic structure as Eurasian and African plants or animals without having a common ancestor with them is nil. There are just too many possible forms of life for each to occur more than once. How, then, would the same varieties of life exist in two such isolated continental environments?

The Americas on Lunholm would have many of the same plants and animals as Eurasia and Africa because all these continents were once attached, like their counterparts on Earth. As discussed in the previous chapter, continents are the highest regions of tectonic plates that move continually over the surfaces of Earth and Lunholm. Geologists have matched similar rock on different continents, thereby connecting these bodies in a giant jigsaw puzzle. A quarter of a billion years ago all the continents on Earth were attached to one another. On this supercontinent, called Pangaea, the animals and planets spread onto what would become the American continents. When the continents parted, they carried many of the same species.

Knowledge of Evolution

While individual species had unique origins, the differences between subspecies on different continents developed through continued evolution. Once a species moves into a new habitat, the animals often adapt to it by evolving specialized characteristics to make them more competitive in their new home. On Earth this concept was codified by Charles Darwin and his successors following Darwin's epic-making 1831 voyage to the Galapagos Islands in the Pacific Ocean. There he found that several species of birds had evolved different characteristics on different islands. Since Darwin would never have made that trip on Lunholm, his discovery of the evidence for evolution would have to be found elsewhere.

Even without visiting the Galapagos, evidence of evolution could still be found on Lunholm. There are always places on single continental landmasses where a species has been isolated from its kin long enough to develop distinct traits. For example, squirrels living on the north rim of the Grand Canyon are clearly related to those living on the south rim, but they also have differences.

The problem with making discoveries about similar species on the same continent, rather than on islands, is that members of the same species differentiate precisely because they are isolated from one another. Specialized sub-species are often found in remote and inaccessible valley or mountain habitats. Because of their isolation, getting to these locations is difficult. The proof of evolution on Lunholm might remain hidden until well into the era of flight, when travel to remote areas would be done in helicopters. Other sciences besides the science of evolution would develop more slowly on Lunholm than they did on Earth. In particular, astronomical observations would be much harder to make there in the bright glow of the closer moon.

Astronomy

On Earth detailed observational astronomy began in the sixteenth century. Tycho Brahe, the premier observer of the time, made meticulous observations of the positions of stars and planets that led to Johannes Kepler's discovery of the nature of planetary orbits in the early seventeenth century.

On Lunholm, with the moon casting sixteen times more light on the planet than our moon casts on Earth, the bright nights would make astronomical observations such as this much more difficult. Most of the light from stars now visible would be washed out by the blue sky that would glow during most nights. Only when Lunholm's closer moon is "new," or a fine crescent, would the sky be dark enough to show all the stars visible today. At other times only a few hundred of the brightest stars would be visible. Therefore, Lunholm's early astronomers would have fewer stars to use as references as they tried to understand the motion of bodies in the sky. Indeed, it might take an extra century or two until someone verified Copernicus's theory of the sun-centered solar system.

On Lunholm the discovery of the sun-centered solar system would come in the equivalent of the nineteenth century, rather than the seventeenth century. The long delay in placing the earth in its rightful orbit around the sun would allow the greater growth of nonscientific theories of the earth and universe. Since observations of sidereal orbital periods of planets and moons help test physical theories, this delay would also hinder the emergence of the science of physics as we know it today.

Clouding Lunholm's atmosphere with light, the closer moon would prevent astronomers from seeing much of the heavens. Until astronomical observatories were placed into planetary orbit above the atmosphere, astronomers on Lunholm literally would be groping in the dark for clues about what the universe contains, how its stars and galaxies work, and how it is evolving.

In these first two chapters the moon has been the arbiter of life on Earth. Two other bodies play essential roles in that regard: the earth itself and the sun. The earth could be bigger or smaller. It could be rotating faster or slower, or its axis could be pointing in another direction. It could have a different chemical composition. It could be located at a different distance from the sun.

The sun could be bigger or smaller, younger or older. It, too, could have a different chemical composition. Some of these changes would prevent life from existing on Earth. For example, if the earth were slightly closer or slightly farther away from our present sun, it would be respectively too hot or too cold to support life. If the surface of the earth contained significantly more metal than it does, the oxygen returned to the air from the oceans might have all combined with the iron. As a result there would not be enough oxygen left in the air for land animals to breathe, preventing surface creatures from evolving here.

In the chapters that follow we will explore changes to both the earth and sun that are consistent with life here. Returning the moon to its present distance, we will next consider what life would be like on a smaller earth.

WHAT IF THE EARTH HAD LESS MASS?
PETIEL

AMONG THE INNUMERABLE FACTS OF LIFE THAT WE ALL TAKE completely for granted, gravity is paramount. Although we don't always weigh what we would like to, we never question the fact that we have weight. Indeed, Ben Franklin, when he said "In this world nothing is certain but death and taxes," might have included gravity as one of life's certainties. In this chapter we will explore the implications of the earth's gravity on life and other natural processes.

From birth until death, gravity continually challenges each of us. As soon as we are born, the earth wraps its gravitational cloak around us and dares us to arise. Slowly our minds learn what our muscles must do in order for us to move. Overcoming the first gravitational bonds, we learn to crawl and then to walk. As toddlers we all paid the price for accepting gravity's challenge with innumerable bumps and bruises. But our competence at walking grew, and in time we ran, jumped, and leapt, receiving even more bruises in return. After years of practice we finally master gravity, at least as far as our bodies are designed to. But until very recently, gravity has still kept us in thrall on the earth's surface.

For millennia humans watched birds soar gracefully through the air. They saw our feathered kin exuberantly wheel and turn, float and dart. Birds see the world from a viewpoint unavailable to humans, and many of our ancestors envied them their freedom and exalted vantage of the earth. The desire to fly probably dates from the earliest humans, but only in this century was the dream fulfilled. First with mechanical aids and then with leg power alone, humans released one more surly bond of gravity and soared into the air. The thrill of flight leaves a mark on everyone who experiences it, but even flying wasn't enough for some. The desire to break the last bond, to completely escape Earth's gravitational pull, helped propel us into the space age and voyages to the moon.

Adept in using our bodies to travel on the earth's surface, complacent in taking intercontinental flights, most adults rarely think about the earth's gravitational pull on us. In this chapter we will examine gravity's effects on us, by imagining a world with less of it. How would lower gravity affect the rest of the natural world? How would our lives and those of the other creatures on Earth be different if the planet pulled us down with less force? To begin to explore these questions, we need to understand exactly what is meant by gravitational force and how it would be possible to reduce it.

GRAVITY

Gravity (or gravitational force) is the mutual attraction between all pairs of objects in the universe. For example, the earth pulls everything on its surface down. It also pulls the moon toward it. In return, the moon's gravity pulls on the earth with equal force. The strength of this mutual gravitational force is so strong that the earth and moon are locked forever in a waltz around the sun. Clearly the sun pulls on the earth and all its inhabitants, but so too do all the billions of other stars in our Milky Way galaxy. As a result, the sun and planets parade around the center of the galaxy completing one circuit every quarter of a billion years. All the billions of other galaxies pull on the Milky Way and on all the stars and planets in it. Gravity is unique in being a universally attractive force.

The gravitational force between two objects depends only on their masses and their distance apart. An object's mass is the total number of atoms and molecules it contains. The greater their masses and the closer they are together, the more strongly two things attract each other. The converse is also true. For example, if the earth kept its present size but doubled its mass, it would pull us down with twice as much force; we would weigh twice as much. On the other hand, if the earth had twice its size while retaining its present mass, it would pull on us with one quarter of its present force; we would weigh one quarter what we do now.

Mass and weight are fundamentally different qualities. A body's mass is independent of where it is located. A person has the same mass on the earth, on the moon, or in space. On the other hand, how much an object weighs depends on where it is located. For instance, a person who weighs 150 pounds standing on a beach would weigh only 25 pounds on the moon, which has less mass. That same person would be weightless floating in space. For our purposes, an object's weight is a measure of how much force is exerted on it by the earth or moon.

Even on Earth an object can change its weight without changing mass. Weight depends on where the scale is located. When you stand on a mountaintop, you weigh less than when you stand on a beach. A 150-pound person would weigh only 149½ pounds on the top of Mount Everest. This occurs because on the mountain you are farther from the bulk of the earth: The closer two objects are, the more they attract each other, and vice versa.

MASS, DENSITY, AND RADIUS

Another important quality of matter is how closely packed its atoms and molecules are. Suppose you had a piece of lead weighing exactly as much as you. That lead would take up less volume than you do because the particles making up the lead atoms are packed more tightly together than are the carbon, hydrogen, oxygen, and other atoms in your body. The lead has a higher density.

Density gauges the amount of mass an object has in the volume of space it fills. An object's density is determined both by its chemical composition and by the amount of compression it experiences from the world around it. However, despite having different volumes, your body and an equal weight of lead have the same mass. Keeping in mind these three concepts of mass, weight, and density, we now turn to lowering the earth's gravitational force.

There are several ways to create an Earth with lower gravity. It could be made of lower-density substances while remaining the same size as our Earth; it could be smaller, having smaller amounts of the same materials as are in our Earth now; or it could have both modifications. If less of the earth today were made of dense materials such as iron and nickel and more were made of lower-density silicon and aluminum, the earth would have less mass. Standing on the same-sized but lower-density earth, we would feel less gravitational pull from the lower mass below our feet. Therefore, we would weigh less.

Equivalently, if the earth kept the same density it has today but had less mass than it does now, we would feel less gravitational pull from it. A lower-mass, equal-density Earth would be smaller than our earth. To lower the gravitational force we feel standing on the earth's surface, we must clearly specify both the planet's mass *and* density.

The fact that the lower-mass earth must be able to support life provides us with guidance in fashioning this new world. Life as we know it requires a large number of different chemical elements and compounds. Some, such as carbon, nitrogen, oxygen, hydrogen, and calcium, are needed in large quantities. Others such as selenium, tin, zinc, copper, and iodine, are needed in minute quantities. In any event, we humans require twenty-five chemical elements in order to function as we do.

Changing the lower-mass earth's density explicitly means changing its chemical composition. Any chemical change in the world changes the amounts of each chemical available for use by living organisms. With our present limited knowledge of biochemistry, it would be impossible to accurately predict all the effects of changing chemical concentrations. Could life function effectively with less selenium or more zinc? What changes in activities, appearances, and abilities would such chemical modifications bring about?

Lacking detailed answers to such questions, we choose to leave the earth's chemical composition alone. That will enable life forms very similar to those found here to evolve. Therefore, the lower-mass earth will have the same density (and hence chemical composition) as our world. We will just shrink the size of lower-mass Earth in order to decrease its gravitational force.

CHOOSING A LOWER-MASS EARTH

To hold its atmosphere in place, the earth must have enough mass to produce sufficient gravitational force. Even today some of the atmosphere is leaking into space. Atmospheric gases that are heated sufficiently by the sun develop enough energy and speed to overcome the pull of gravity and fly away. When we lower the earth's mass and gravitational pull, it will be even easier for the atmosphere to evaporate. If the lower-mass earth's gravitational pull on its atmosphere were too low, virtually all the gases needed for life to evolve would be depleted.

Clearly, then, the earth has to have enough mass to create the gravity needed to keep the heated gases of the air from escaping into space. To maintain a breathable atmosphere, the earth would need to have enough mass to retain oxygen, nitrogen, carbon, water vapor, and carbon dioxide. The oxygen provides the energy needed by animals. The nitrogen serves to dilute the oxygen so that life is not burned alive, as discussed in chapter 1. The water vapor must not escape, so that it can close the cycle of evaporation from oceans and condensation as rain and snow. In other words, the water that gets evaporated from oceans and seas eventually needs to return to the earth. If this gas drifted into space, the earth would eventually lose its water supply and life would cease to exist. The carbon dioxide, which is only $\frac{3}{100}$ of a percent of the atmospheric gas, provides food for plant life and historically was the source of the oxygen in the air today.

The speed that gas molecules need in order to depart into space, called their escape velocity, is the same for all gases, and depends only on the mass and radius of the earth. The escape velocity from the present earth is around sixteen thousand miles per hour. This far exceeds the typical speeds of all the gases listed above, which is why the earth has maintained its present atmosphere for billions of years.

Lowering the earth's mass reduces the escape velocity from it. The smallest mass that the earth could have and still retain all the gases necessary for life is one quarter of its present mass. The corresponding radius of the smaller earth would then be 2,500 miles, which is just over 60 percent of its present radius of nearly 4,000 miles. This is the mass we choose for the lower-gravity earth.

We call this petite version of Earth Petiel. In these respects Petiel is like Earth: Its day is twenty-four hours long, it is the same distance from the sun as Earth, it has the same chemicals in the same proportions as the earth, and it has an identical moon orbiting at the same distance as our moon. Petiel's continents and oceans would all be smaller than they are on Earth, and their locations would depend on how the planet's surface evolves.

PETIEL'S ASTRONOMICAL PROFILE

Petiel's diameter would be only two-thirds that of the earth. Petiel would still be larger than Mars, which is only half the size of the earth. Everything on Petiel would weigh two-thirds as much as it does now. Perhaps contrary to intuition, Petiel's lower mass would have no effect on the rate that it orbits around the sun (the length of a year). Planetary orbital periods around the sun depend only on the mass of the sun and the distance from the planet to the sun. Petiel, at the same distance from the sun as we are today, would orbit the sun at the same rate that the earth does. Indeed, objects of any mass orbiting the sun at the same distance as the earth take exactly one year to go around.

However, the orbit of the moon around Petiel would be very different from the orbit of the moon around the earth. The orbit of a moon is determined by the mass of the planet it orbits and its distance. Petiel's lower mass would increase the moon's sidereal and synodic periods around the earth (see chapter 2, pages 66–67, for definitions of these periods). The moon would orbit Petiel in fifty-four and two-thirds days. A lunar cycle of phases would take fifty-nine days. Since it would take the moon longer to orbit Petiel, there would be fewer new moons and full moons there than here.

As a result, on Petiel there would be fewer chances for eclipses to occur each year than there are on Earth.

TIDES

We saw in chapter 2 that Lunholm's closer moon generated higher tides than does our moon today. Conversely, the moon orbiting smaller Petiel at the same distance as our moon would create lower tides on that planet. There are two reasons for this. First, tides decrease with the diameter of the planet. The smaller the planet is, the smaller the tides created on it by its moon or by the sun. Since Petiel's radius is only two-thirds that of the earth, the tides on it would only be two-thirds as great as our tides.

Second, tides would decrease if the moon were moving more rapidly across the sky than our moon does. This is just the situation on Petiel, despite the fact that the moon takes longer to orbit the planet. To understand why, imagine that you are sitting in a train traveling at one hundred miles an hour parallel to a highway. If the cars on the highway travel at sixty-five miles an hour, your train will slowly pull ahead of them. If the cars travel at thirty miles an hour, your train rushes past them. The earth and moon are represented by the first situation, where the moon is orbiting the earth relatively rapidly, compared with the rate that it orbits Petiel. Since the moon orbits the earth in the same direction that the earth rotates, the earth slowly "pulls ahead" of the moon. The case of the slower cars is analogous to Petiel and its moon. That moon orbits more slowly, so if you were standing on Petiel, the moon would appear to move quickly across the sky just as the slower cars move quickly by.

Since the oceans on Petiel would be "under" the moon for less time than they are on Earth, the water would have less chance to rise toward the moon before being pulled away by Petiel's rotation. The rapid motion of Petiel's moon across the sky is just the opposite of what would be experienced on Lunholm in chapter 2. There the closer moon would orbit more rapidly than our moon and thereby stay above each part of the planet longer. Lunholm's moon would move across the sky of that planet more slowly than does the moon of either Earth or Petiel. In summary, the tides on Petiel would always be lower than those on Earth and Lunholm but greater than those on moonless Solon.

INTERNAL HEATING

Differences in volcanic, seismic, magnetic, and thermal activities on Earth and Petiel would be determined primarily by the differences in the planets' interiors. Both would have molten interiors, although Petiel would have considerably less magma than there is inside the earth.

It may seem odd at first glance that the interior of a planet, which is not exposed to the heat of the sun, is hotter than its surface. This occurs for several reasons, starting with the formation of the planet. For millions of years after they form, planets are kept molten by infalling space debris. The surface radiates heat into space after the impacts subside, thereby cooling and solidifying into rock. Thereafter the surface does not generate any more heat of its own; it merely radiates heat it receives from the sun and from the planet's interior.

The impact energy that makes a planet's surface molten also liquifies its interior. Since there is much more mass inside the planet than on its surface, the interior holds more heat than does the surface. For example, there is fifty times more mass in the earth's interior than on its surface (its crust). Therefore it takes longer for a planet's interior to shed its heat and solidify. The cooling of the interior is slowed further, compared with the cooling of the surface, because interior heat must travel through the solid crust. The crust serves to insulate the interior from the cold of space, since solid rock is a very bad conductor of heat. Therefore, heat stored in a planet's interior leaks out slowly, over billions of years.

Planetary interiors also have at least two sources that replenish some of the heat lost through the surface. These are radioactivity and compression. Radioactive elements are those that spontaneously break apart into two smaller elements. This separation is accompanied by the emission of a considerable amount of heat, which rewarms the interior. Sufficient concentrations of radioactive elements supply enough heat to melt rock (or to keep it molten), as was seen so graphically at the meltdown of the Chernobyl nuclear reactor.

Chernobyl's reactor contained concentrated radioactive elements, which lost their coolant. Without its heat being removed, this radioactive material overheated and melted the vessel that contained it. In the same way, radioactive elements such as uranium and plutonium persist inside a planet for billions of years, helping to keep its interior molten.

Compression by a planet's outer layers also heats its interior. Indeed, anything that is compressed heats up. However, because the interiors of Petiel and the earth are made of relatively incompressible rock and metal, this process shrinks these planets relatively little and generates much less heat than does radioactivity.

We can now see why Petiel's interior would have less magma than the earth does. In the first place, Petiel's surface area is 45 percent as great as the earth's surface, while its interior contains only 30 percent as much mass as the earth's. Therefore, Petiel's surface is larger *relative to that planet's volume* than the earth's surface is in relation to its volume. Having relatively more area to leak through, Petiel's surface would radiate its interior's heat more quickly than occurs on Earth.

In the second place, Petiel's radioactive elements would provide less heat to the interior than do those elements in the earth. Although Petiel would contain the same fraction of radioactive elements as does the earth, its smaller size means that overall it would have less total radioactive matter. Since most radioactive elements in a planet collect in its central regions, the earth would have more concentrated radioactivity than would Petiel. Therefore, Petiel's core would generate less of its own heat than does the earth's core. As a result, more of Petiel's interior would be solid.

Indirect evidence in support of the idea that Petiel would be more solid than Earth comes from the two planets in our solar system most like Petiel: Mars and Mercury. Mars has 10 percent and Mercury has 6 percent as much mass as the earth; they have 40 percent and 24 percent as much mass as Petiel, respectively. All indications are that Mars and Mercury, which are similar to the earth in chemical composition, contain considerably less magma than the earth does. Petiel, falling between the earth and these other planets in mass, would presumably have an intermediate amount of heat and magma as well.

As a result of more efficient cooling and less postformation heating, Petiel's solid crust would extend deeper into the planet than Earth's crust does into Earth. Magma from deep inside Petiel would find fewer cracks and weak spots through which to push to the surface.

TECTONIC PLATES AND PETIELQUAKES

Petiel's mantle, the layer of rock just below the crust, would be cooler and stiffer than the earth's mantle because of Petiel's greater heat loss. The earth's heat keeps its mantle rock plastic, meaning it is easily deformed by magma pushing up through it. Petiel's mantle rock would be more rigid and therefore more resistant to the flow of magma through it. On Petiel magma would have a harder time penetrating the mantle to the crust than does magma inside the earth. Likewise, Petiel's stiffer mantle would also resist the crust sliding over it. Therefore magma breaking through the crust would have greater difficulty pushing aside the tectonic plates carrying the continents.

In fact, it is likely that long before Petiel's present age of 4.6 billion years, its mantle would have become so thick and unyielding that even when lava did occasionally break through the crust, the surface would no longer move in response. This also seems to be the case with Mars and Mercury, which do not show tectonic activity, while Earth and Venus (with 82 percent of the earth's mass) still have tectonic-plate motion.

Plausibly, Petiel's tectonic plates stopped moving relative to one another 3½ billion years after the planet formed, roughly a billion years ago. Since then, Petiel's continents would have been fixed in position. We assume that Petiel's continents were connected at the time its tectonic plates froze in place. While it would certainly have islands, the bulk of its landmass would be one vast supercontinent. This is consistent with the fact that the earth's continents were once all attached.

Repositioning the continents would have a major effect on Petiel's global weather. Climate is strongly affected by the location of oceans and continents, as well as by ocean currents, wind patterns, and many other factors. The differences between the locations of the continents on Petiel and on Earth—and consequently the difference in ocean currents—would make the weather patterns on the two planets very different.

Petiel's thicker crust would also greatly diminish quake activity compared with that on Earth. On Petiel, boundary regions between tectonic plates, such as the San Andreas Fault on Earth, would be permanently fixed and relatively free from petielquake activity. Whereas on Earth islands along the east coast of Asia, including Japan and the Philippines, often endure large earthquakes, on Petiel they would be seismically inactive. These islands, and the Hawaiian islands, which all have active volcanoes, would have little volcanic activity if they were on Petiel. In general, both inside and out, Petiel would be more tranquil than the earth, with fewer volcanoes and quakes.

VOLCANOES

Although volcanos would be rare on Petiel, those that do erupt would do so explosively, with a blast more spectacular than any experienced on Earth. These explosive volcanoes are called stratovolcanoes, and their eruptions are among the most dramatic natural events on Earth. They eject debris miles into the air. Stratovolcanoes occur when the magma moving up from inside the earth encounters relatively thick, dense rock near the surface. The magma builds up pressure and blows the rock off as if it were a cork popping off a champagne bottle. After the initial eruption of a stratovolcano, lava slowly leaks out. Because of their power and the shapes of the mountains they form, stratovolcanoes are very exotic. They include Mount Fuji in Japan, Crater Lake and Mount Hood in Oregon, and Mount Rainier and Mount Baker in Washington.

Less eruptive volcanoes, from which lava flows relatively easily from beginning to end, are called shield volcanoes. They exist where the earth's crust does not strongly prevent the magma from emerging. The molten rock from inside can push through to the surface of the earth in such volcanoes without having to explode the crust. Mauna Loa and Mauna Kea on Hawaii are examples of such volcanoes, as are volcanoes on Tahiti, Samoa, and the Galapagos.

Petiel's cooler, firmer mantle would serve to cap all upwelling magma. Because magma nearing its surface would have to push very hard to move the rock over it, all volcanoes on Petiel would be stratovolcanoes. Their spectacular eruptions would be due in part to Petiel's magma buildup and in part to its lower gravity, enabling the volcanic ejecta to soar higher in the air after the eruption. Once launched by the magma, the volcanic matter would soar 60 percent higher than it does on Earth. When Mount Saint Helens erupted in 1980, it blasted over 5 billion tons of gas and dust up to six miles into the air. On Petiel, the same explosion would have sent this matter up to ten miles high.

The altitude to which volcanic debris climbs is very significant in the effects this material has on a planet's temperature and habitability. Because it would shoot higher, the gas and dust from volcanoes on Petiel would be spread over larger areas of that planet's surface by high-altitude winds. Being higher up, this material would also take longer to fall back to the surface than it would on Earth.

While in the air, the dust would reflect light and heat from the sun back into space. That sunlight would not reach Petiel's surface. Therefore the days would be darker and cooler following volcanic eruptions on Petiel until the air cleared some months or even years later. The same happens on Earth following volcanic eruptions, but since the dust goes up to a relatively low altitude, it falls out relatively quickly. Dirt in the air makes for colorful sunsets and sunrises. After volcanic eruptions these times of day would be even more spectacular on Petiel than they are here.

More important to life than the dust from Petiel's volcanoes is the sulfur dioxide gas they would be injecting into the atmosphere. From our experience on Earth, we know that the sulfur dioxide on Petiel would mix with water molecules to become sulfuric acid droplets. Because of the extreme altitude to which Petiel's eruptive gases would soar, the sulfuric acid they create would remain aloft for years, perhaps even decades. There the acid would absorb some sunlight, darkening the planet further. Eventually the acid would fall to the ground as acid rain.

While all the dynamics of volcanoes on Earth affect the climate and, to a lesser extent, the soil, they have clearly not prevented life from evolving. If volcanoes on Petiel erupted with the same frequency as they do on Earth, however, we would indeed expect them to have an adverse effect on life there. Volcanic emissions would keep the seasons colder and make the soil more acidic than Earth's. Happily, as we have seen, volcanoes should erupt much less frequently than on Earth. Given time for the air and soil to recover, relatively infrequent volcanic eruptions should not prevent complex surface life from evolving there, nor should they wipe out that life after it forms.

AIR

Since it would have one quarter the mass of Earth, Petiel would contain less rock in its interior from which carbon dioxide gas could be released. Consequently, Petiel's early, carbon dioxide–dominated atmosphere would have only one quarter as much gas as did the earth's early atmosphere.

By an interesting twist of geometry, on Petiel a larger fraction of atmospheric carbon dioxide would be absorbed in the oceans than was absorbed on Earth. That would work as follows: Petiel would have 45 percent as much surface area as does the earth. Therefore, its ocean volume would be 45 percent of that on Earth. Since Petiel's rocks would give off 25 percent as much carbon dioxide as Earth's rocks did, Petiel's oceans would have much less carbon dioxide to absorb from the air than did Earth's oceans. So, even though Petiel's oceans would be smaller than the earth's, they would be able to absorb a larger fraction of the atmospheric carbon dioxide than did Earth's early oceans.

Here is an analogy that might help clarify this process. Consider two bowls, one containing 25 percent as much water as the other. In the bowl with the most water, put a sponge that can absorb half the water. In the bowl with the least water, put a sponge that has only 45 percent as much volume as the first sponge. The smaller sponge, having less total water to absorb, will collect 90 percent of the water in its bowl. While the bigger sponge will absorb more total water than the smaller sponge, the smaller one will actually absorb a larger percentage of the smaller volume of water in its bowl. The sponges represent the oceans; the water represents the atmospheric carbon dioxide.

Petiel's ocean life would have less total carbon dioxide to convert to oxygen. Therefore the concentration of oxygen in Petiel's final atmosphere would be lower than it is in the air we breathe. As on Earth, the first oxygen returned to the atmosphere would be absorbed by other chemicals. Plausibly, the final oxygen level on Petiel would be as high as one quarter the oxygen content of our air. As we'll see shortly, the lower density of oxygen in Petiel's atmosphere would have important physiological effects on animals.

Turning now to the meteorological effects of the lower-density atmosphere, recall that winds are created by a combination of heating from the sun and planetary rotation. Petiel rotates at the same rate as the earth. The heating of its lower-density air would generate higher winds than occur here. We see this same phenomenon on Mars, which rotates at the same rate as the earth and which has an atmosphere $\frac{1}{50}$ as dense as ours. Storms on Mars are often planetwide, with winds blowing over 150 miles an hour. These storms sometimes last for weeks.

Compared with those of the earth, the lower atmospheric densities of Mars and Petiel make it easier for the winds to move rapidly because the air molecules bump into fewer other molecules as they travel. This lowering of the air density has the same effect on winds as increasing the rotation rate of the planet, as we saw in chapter 2.

METEORITES

Petiel's thinner atmosphere would be less effective in burning up infalling space debris than is the thicker atmosphere of the earth. Such debris enters the earth's atmosphere continuously. Heated by air friction as they plummet through the atmosphere, these pieces of rock and metal begin vaporizing their outer layers. This vapor creates a trail across the sky, which you can see on any clear night. We call these streaks meteors or shooting stars.

Most of the incoming meteoritic material today is less than a few inches across and virtually all of it is less than a few yards across. (Since the much larger pieces of debris that cause mass extinctions of plants and animals are extremely rare, we are not considering them here.) Most meteors fail to reach the earth's surface intact; they burn up completely in the atmosphere. Those meteor remnants larger than dust particles that do reach the ground are called meteorites. Having lost most of their mass and energy while passing through the air, meteorites rarely do any harm.

The same could not be said about meteors falling through Petiel's thinner atmosphere, which would generate less friction to heat the space debris. With less heat vaporizing less of their surfaces, shooting stars would have shorter and dimmer tails. Since fewer meteors would vaporize completely before they struck the ground, Petiel would be pelted more frequently with intact meteorites than is the earth.

While individual meteors are small and of low mass, there are enormous numbers of them each day. By comparison to the number of meteoroids falling toward the earth, roughly ten thousand meteoroids would enter Petiel's atmosphere daily. Many more of these would strike its surface intact.

RIVER FLOW

The lower gravitational force that Petiel exerts on its surface would have far-reaching consequences on the flow of rivers to the sea. Rivers flow downhill, finding their way to the lowest places on a planet's surface. Lakes and seas form where the surface of the earth (or Petiel) has a local depression. These bodies of water fill until they overflow a bank somewhere. Water thus spilled continues downward until it empties into an ocean.

The rate that a river flows depends on how steeply it descends, on its width and depth, on the kind of material over which it flows, and on the amount of dirt and other substances, called silt, it carries. Some silt settles to river bottoms, eventually filling them and changing the courses the rivers follow. The more slowly rivers flow, the more silt they drop onto their riverbeds. Deposited silt is called sediment. Silt that doesn't settle is carried to the ocean and is either deposited at the mouth of the river or dispersed by tides and waves.

Because the gravitational force pulling river water downstream would be lower on Petiel than it is on Earth, rivers on Petiel would flow more slowly than they do here. As a result, the silt in them would have more time to form sediment on river bottoms than does silt in the more rapidly moving rivers of Earth. More sediment on river bottoms means less sediment deposited in the ocean. Rivers would change course more frequently on Petiel, and more importantly for the early evolution of life, fewer minerals necessary for the formation of early life would be deposited in Petiel's oceans.

LIFE ON PETIEL

Lower tides and less silt in the oceans mean that it would take Petiel's oceans longer than Earth's to accumulate the chemical building blocks of life. In all likelihood, however, once the oceans amassed enough material, life would evolve. The later emergence of animal life from Petiel's oceans onto its shores would be facilitated by the lower tides there, allowing early amphibians safe, easy access to the land.

Immobile Continents and the Diversity of Life

Long before life would make the transition from Petiel's oceans to its surface, the tectonic plates there would have stopped moving and the continents, or rather the supercontinent, would be fixed in place. As a result, the rate with which species would diversify and spread would be different on Petiel than it was on Earth, where continents separated over time and created new niches for life.

We noted in chapter 2 how life forms in different milieus, such as different continents and islands on Earth, tend to evolve differently. They respond to the specialized needs of their immediate surroundings. This is especially true when the range of a given species becomes geographically isolated into several regions. The ability of animals and plants to range over virtually all of Petiel's landmass would cut down on the number of distinct species that would evolve there. The diversity of both plant and animal life should therefore be less on Petiel than it is on Earth.

Animal Sizes

The size of animals on Earth provides our best insight into how big creatures on lower-mass Petiel might be. Earth's creatures range in weight from over eighty tons to less than an ounce. Multiton dinosaurs were at the top of Earth's food chain for tens of millions of years, while insects such as cockroaches, weighing only ounces, have survived successfully for even longer. Apparently nothing on Earth sets an a priori ideal size or weight. Therefore, we can expect animals on Petiel to also have a large range of sizes and weights.

It is nevertheless likely that Petiel's lower gravity would have some noticeable effects on animal life there. Lower gravity means animals of any mass would weigh less than they do on Earth. To move a given mass would therefore require less musculature, all other things being equal. Less weight requires less structural strength to hold it up, so bones could be thinner than they are here and still support animals of the same mass.

All these comparisons between animals on Petiel and on Earth assume that the only differences in the creatures' physiologies come directly from the difference in planetary mass. There would, however, be many other factors indirectly affecting evolution that result from Petiel's lower mass. These include the different winds, the different distribution of continents and oceans, the different climates created by different weather patterns, and the different amount of oxygen, among many others. Let us consider how one factor, the effect of Petiel's lower oxygen concentration, would change the animals there.

Effects of Diminished Oxygen on Petiel's Life

Oxygen plays a vital role in enabling animals to convert their food into useable energy. Nature has evolved ways to create energy biologically without using oxygen. Called anaerobic metabolism, these biochemical mechanisms are incredibly inefficient compared to oxygen-consuming (aerobic) metabolism. For example, the efficiency of aerobic metabolism is nearly twenty times greater in forming adenosine triphosphate, ATP, than is anaerobic metabolism. ATP is the major biochemical energy source in virtually all forms of life.

Life forms based entirely on anaerobic metabolism would make much less energy and many more undesirable chemical by-products than does oxygen-consuming life. It is extremely unlikely that life based on anaerobic biochemistry could evolve into the diverse, complex life that inhabits the earth today.

The amount of activity shown by different life forms on Earth depends in part on the amount of oxygen they consume for each ounce of body tissue. To be sure, many, many other factors are involved in animal metabolism, but this relationship is still important. Which, if any, animals would Petiel's lower oxygen level hurt most?

On earth metabolism rates vary tremendously. They are highest among the smallest animals and lowest among the largest animals. This happens for two reasons. First, smaller animals radiate heat more rapidly than do larger animals, whose bulk helps them store heat. This is the same concept we used to explain why Petiel would be cooler than the earth; smaller Petiel would have a larger surface area relative to its volume than does the earth. The same is true for small animals compared with larger ones (this applies especially to fat, round animals such as mice and elephants). Therefore, to create enough energy keep their body temperatures at operating levels, smaller animals must eat more and breathe in more oxygen in relation to body mass than do larger animals. Second, smaller animals need to move faster than larger animals in order to find food and evade larger predators. "Faster" in this context is measured in terms of body lengths traveled per minute rather than miles per hour.

Smaller animals consume a larger amount of the energy they create each second than do larger animals. For example, a mouse weighing seven-tenths of an ounce consumes nearly six times more oxygen per ounce of body weight than a cat does. A mouse consumes over eleven times more oxygen per ounce of body weight than a person does and over thirteen times more than does a cow. Active small animals, such as hummingbirds, consume even more oxygen per body weight than do more sedentary animals of equivalent mass. Relatively high oxygen-consumption rates are essential for small animals if they are to compete successfully with larger, stronger animals.

Animals on Petiel would have less oxygen in the air they breathe than do animals on Earth. Since anaerobic metabolism would be far too inefficient to support complex life, animals there would have to evolve adaptations to the lower oxygen level. For the sake of similarity to life on Earth, let us assume that Petiel's animals evolve so as to consume as much oxygen as does life here. In order to do so, Petielians would have to breathe four times faster or have lungs with four times as much surface area (or a combination that amounts to four times more respiration) as their terrestrial counterparts.

The creatures on Petiel who would lose most as a result of their atmosphere's lower oxygen content are the very small ones. If they remain small, they would not be able to have larger lungs to absorb sufficient oxygen. Therefore they would have to breathe four times faster. But this immediately throws them into a catch-22 dilemma because the process of breathing takes energy.

Breathing faster would require the small creatures to use more muscle power, which in turn would require them to consume more oxygen. This would require them to breathe faster still, which would require them to use more muscle power, and so forth. Either the very smallest animals on Petiel would need larger lungs, making them bigger creatures, or they would have to function with less oxygen than they do on Earth. However, using less energy would slow the small animals down, making them more vulnerable to predators and less able to attack prey. Lower oxygen levels on Petiel could therefore have the effect of limiting the activity and even existence of very small life.

The most energy-efficient way to evolve animals on Petiel with similar oxygen-consuming capacity to that of animals on Earth is to give them larger lungs or other oxygen-collecting surfaces. This would allow them to breathe at the same rates as animals here, thereby minimizing the amount of extra energy they would need. Nevertheless, animals on Petiel would have to work harder to obtain the oxygen they require.

Larger lungs would imply larger chest cavities. Animals would have to be physically larger and their lung muscles stronger than they are here. The increased mass of the scaled-up animals on Petiel would require more massive skeletons to support them. In that way, the lower oxygen content in Petiel's air would partially offset the decreased bone and muscle mass allowed there by the planet's lower gravity.

Birds

The lower gravity on Petiel might seem to imply that birds with the same muscle mass as birds on Earth would have an easier time flying. Even taking into account their larger lungs, it would seem that birds could weigh less on Petiel than they do on Earth. It might also seem that their muscles would have to work less to lift them.

Unfortunately for birds there, Petiel's lower air density means that bird wings would have less lift than they do on Earth. Wings would have to beat harder or be larger in order to achieve and sustain flight. In either event, birds would require more muscle power, respiration, and weight than is required by their counterparts here on Earth.

Humans on Petiel

We will first assume that humans on Petiel evolve to consume as much oxygen as we do so that their brains and bodies can function at the same rates as ours. To accomplish this they would need either to have larger lungs or to breathe faster, as we discussed earlier. Besides requiring more exertion, more muscle mass, and the consumption of more energy, faster respiration can limit how completely gases are exchanged in the lungs. In other words, breathing too rapidly prevents inhaled oxygen from being as completely taken into the bloodstream in the lungs as do longer, slower breaths.

It seems reasonable, therefore, that like other animals on Petiel, humans there would collect oxygen primarily by having larger lung capacity than we do. Assuming they are erect primates, they would need to be either taller or more barrel-chested than we are in order to accommodate their larger lungs. Recalling that Petiel would be a very windy planet because of its lower air density and pressure, it is likely that the primates there would need strong muscles on both their hands and feet to cope with unstable trees. From an evolutionary standpoint, being taller and more lithe would seem to have advantages for the tree-dwelling ancestors of Petiel's humans.

Long bodies, combined with powerful arm and leg muscles, would enable Petiel's primates to jump from tree to tree most easily. Being slender, they would also have better balance than if they were top-heavy, with more massive chests. Humans on Petiel would therefore be taller than we are. They would take deeper breaths to fill their larger lungs. Indeed, such body shapes seem to be the rule among tree-dwelling primates on Earth today.

How would the lives of humans on Petiel be different from ours? Petiel's people would be able to throw farther and higher than we can, since their muscles would launch objects with the same speed ours do while gravity there would slow the projectiles down less quickly. For the same reason, Petiel's humans could jump higher. If they weigh less than we do, they would be able to run faster than us. Their bodies would also feel less wind resistance from the thinner air through which they move.

Petiel's thinner air and lower gravitational force would affect all forms of travel. Consider first the effects on flight. The thinner air would provide less lift on aircraft wings and helicopter rotors. The lower oxygen level in the air would also require aircraft (and land-vehicle) engines to suck in more air in order to burn their fuel completely. On the other hand, the lower air density would provide less drag on aircraft, so Petiel's planes would have many different design characteristics than ours do. Getting into space would be easier from Petiel than it is from Earth because of the lower gravitational force there. Rockets would require less thrust to lift their loads and to escape from Petiel's gravity.

Finally, suppose that humans on Petiel evolve from arboreal animals that consume *less* oxygen and therefore have less mental functioning than did our ancestors. Could such slower-witted animals survive swinging from tree to tree? One ability essential for the success of tree-dwelling animals on Earth is their rapid response to falling. Because the earth's gravitational force accelerates them downward so rapidly, such animals here have to stop their descent within half a second or face injury or death. Fast reflexes require swift coordination between several parts of the brain and body, which in turn uses lots of oxygen.

On Petiel, the force of gravity would only be 63 percent as great as it is on Earth. Tree-dwelling animals with slower reflexes would therefore have that much more time to react before falling dangerous distances. In fact, it may well be that the relationship between the acceleration of gravity and the ability of animals to withstand impacts after falling determines, in part, just how responsive brain-body connections have to be. This would help set a lower limit to how much oxygen Petiel's arboreal animals must consume. This, in turn, would set other restrictions on their overall physiology.

WHAT IF THE EARTH WERE TILTED LIKE URANUS?
URANIA

VIVID YELLOW AND ORANGE MAPLE TREES PROCLAIM FALL IN NEW England. The ground is carpeted with a fresh layer of decomposing leaves. The air is sharp and clear, and the redolent aromas of the transforming biosphere herald the coming winter. Farther south the changes are less dramatic, but everywhere in the Northern Hemisphere plants, animals, and people are adjusting to colder days. The hours of daylight dwindle until it seems that if a day had any less light, it wouldn't be worth getting up in the morning. The forests and meadows are discomfortingly quiet; the exuberance of summer is gone. Finally the hours of daylight begin to increase and the days grow warmer. For most people spring and summer come none too soon.

As we will see in this chapter, these changes of season to which we are accustomed are not inevitable. By changing the direction that the earth's axis points as we orbit the sun, it would be possible for every place on Earth to have essentially constant temperatures and weather conditions year-round. A different tilt of the axis would produce seasons that are far more variable than they are now, with all regions of the earth, including the poles and equator, experiencing both snow and tropical heat at different times of the year.

This latter case is particularly interesting because it takes the earth beyond the cycle of longer and shorter days that accompanies seasonal change today into a completely new realm of light and dark. In this chapter we explore what the earth would be like if the seasons were as extreme as possible. Although many people don't like the changes in temperature and precipitation that accompany our seasons, we will see that things could be much worse than they are.

EARTH'S ORBIT AROUND THE SUN

In order to change the cycle of seasons, we need to explore what causes them in the first place. At first glance it seems plausible that the seasons occur because of changes in the earth's distance from the sun. Indeed, the earth's nearest approach to the sun (perihelion) brings it 3 million miles closer than its farthest retreat (aphelion). However, we are farthest from the sun on July 3 and closest to the sun on January 3 of each year. Since the sun is farther away from us during our summer and closer during our winter, something else is clearly causing the seasons. That something is the tilt of the earth's axis of rotation in relation to the earth's path around the sun.

The earth spins once a day around an imaginary axis passing through its North and South poles. This axis is tilted by twenty-three and a half degrees from being perpendicular to the plane in which the earth orbits the sun, called the ecliptic. The name ecliptic is also given to the annual path of the sun among the stars as seen from Earth. The course of the earth around the sun and the "perceived" course of the sun among the stars throughout the year both occur in identical planes. After all, these two paths merely represent the motion of the earth around the sun from two different points of view: the true motion as seen from space and the perceived motion of the sun as seen from the earth. The two uses of the word *ecliptic* are therefore completely interchangeable.

The twenty-three-and-a-half-degree tilt of the earth's axis with respect to a line perpendicular to the ecliptic is called the earth's obliquity. The key to the seasons is the fact that throughout the year both the tilt of the earth's axis (its obliquity) and the direction the axis points in space remain constant. (The gravitational influences of the moon and other bodies actually cause the direction of the axis to change slowly over thousands of years. We ignore these small effects, as they have no bearing on this chapter.) The earth's North Pole points toward the star Polaris in the constellation of Ursa Minor, the Little Bear, throughout the entire year. It might intuitively seem that the axis should change direction as the earth orbits the sun; however, the axis always points in the same direction because of the earth's enormous angular momentum. As mentioned in chapter 2, the greater an object's angular momentum, the harder it is for the object to change direction.

As a result of the earth's axis being fixed in direction, you would always see Polaris directly overhead at night if you stood on the North Pole. Also, standing anywhere else in the Northern Hemisphere and facing toward Polaris, you would always be facing north. All other stars in the sky appear to orbit around Polaris each night because the earth is rotating with Polaris above its pivot point.

If the fixed relationship between the tilt of the earth's rotation axis and the ecliptic is still not clear, you can easily visualize it as follows. Flatten this page and imagine it as the plane of the earth's orbit around the sun (the ecliptic). Hold a pen over the page as you would to write. The pen represents the earth's rotation axis. Keeping your wrist stiff, move the pen over the page in an egg-shaped path to simulate the earth's annual orbit of the sun. You will see that the angle at which you hold the pen above the page remains constant; the pen stays pointed in the same direction throughout the loop, just as the earth's axis always points toward Polaris.

THE SEASONS

Here is how the seasons unfold. On March 22 of each year (plus or minus a day), the sun is directly over the earth's equator. This is one of only two days each year when there are twelve hours of daylight and twelve hours of darkness everywhere on Earth. March 22 is called the vernal equinox ("equal nights"), and it is the first day of spring in the Northern Hemisphere. The autumnal equinox occurs six months later on September 22.

While always pointing toward Polaris, the North Pole points more and more toward the sun as the earth moves in its orbit during the three months following the vernal equinox. You can visualize this by holding a pen in writing position at eye level. Again the pen represents the earth's axis, while your head now represents the sun. The vernal equinox would occur when both ends of the pen are at the same distance from your face.

During the first quarter of an orbit (first quarter of a year) following the vernal equinox, the sun rises higher in the northern sky each day. Therefore it takes the sun longer to go across the sky; the amount of daylight increases daily until finally, on June 22, the earth's North Pole is pointing most directly toward the sun. The North Pole is not pointing *at* the sun, rather it is tilted just 66½ degrees away from it, while the South Pole is pointing farthest from it: 113½ degrees away. This day, when the sun rises highest in the northern sky and creates the greatest amount of daylight ever experienced in the Northern Hemisphere, is called the summer solstice.

On the summer solstice the sun is directly above latitude twenty-three and a half degrees north. At noontime on that day the sun is directly overhead as seen from Havana, Cuba; Muscat, Oman; and Canton, China, among other places. The sun is never directly above any place on earth north of twenty-three and a half degrees north latitude.

This change from the vernal equinox to the summer solstice can also be visualized with the pen. Start with it located in front of your face at the equinox position (top and bottom equal distance from your eyes). The pen, of course, represents the earth and your head, the sun. Move the pen around your head so that its nonwriting end (representing the earth's north pole) is closer to you than its writing point. Keep the pen tilted at a fixed angle to the ground and pointing in the same direction as you move it. You will see that one quarter of the way around your head, the nonwriting end of the pen is tipped most nearly toward you. This is where the axis is at the time of summer solstice.

Starting on June 23, the North Pole slowly begins pointing farther away from the sun as the earth continues in its implacable orbit. As the sun rises to a lower peak in the sky each day, the amount of daylight diminishes in the Northern Hemisphere. This next quarter of a year, from summer solstice to autumnal equinox, is a time reversal of the previous quarter. The autumnal equinox denotes the beginning of autumn in the Northern Hemisphere. Continue your pen's motion around your head (always pointing it in the same direction). The time from the summer solstice to the autumnal equinox brings the top of the pen from pointing most nearly toward you to being at equal distance with the bottom of the pen once again. Unless you have turned with it, the pen is now behind your back.

For the six months following the autumnal equinox, the sun is over different parts of the Southern Hemisphere.

It continues rising lower in the northern sky and higher in the southern sky until, on December 22, it rises lowest in the northern sky. Called the winter solstice by people living in the Northern Hemisphere, this day has the least amount of sunlight in the Northern Hemisphere and the greatest amount of sunlight in the Southern Hemisphere. It represents the beginning of winter for the Northern Hemisphere and the beginning of summer for the Southern Hemisphere. On that day the sun is directly over such places as São Paulo and Rio de Janeiro, Brazil; Walvis Bay, Namibia; and Alice Springs, Australia, all of which are located at roughly twenty-three and a half degrees south latitude. Continuing your pen's progress around you, the bottom point of the pen should be tilted most closely toward your head at the winter solstice.

Finally, the three months from the winter solstice to the vernal equinox bring the sun and your pen back to the starting point of their journeys.

The combination of the sun's changing height in the sky each day and the changing amount of daylight each day is the key to the cycle of seasons. The sun is up higher and for more hours in the northern sky during the six months from vernal to autumnal equinox (spring and summer) than from autumnal to vernal equinox (fall and winter). During the spring and summer, then, the sun has more time to heat the Northern Hemisphere each day than it does during the fall and winter.

Being higher up during the spring and summer, the sun also sends its light and heat straighter down to the Northern Hemisphere's surface than when it is lower in the sky. Therefore, each acre of land receives more energy from the sun during these seasons. This combination of longer days with more intense heat gives us our warmest days during spring and summer in the north. The Southern Hemisphere, of course, goes through exactly the opposite heating and cooling cycle. We turn now to considering what would happen to the seasons if the earth's axis had a different obliquity.

NO OBLIQUITY—NO SEASONS

If the earth's rotation axis were exactly perpendicular to the plane of the ecliptic (obliquity angle of zero), the sun would be over the earth's equator every day of the year. The analogous situation would occur if you hold your pen exactly upright as you bring it around your head in either direction. You can see that unlike the previous case when the pen was tilted in your hand, its top and bottom are now equal distances from your eyes through the entire circuit. Every day would be an equinox with twelve hours of light and twelve hours of dark everywhere on Earth. The sun would always rise to the same height at noon and would provide the same amount of heat to the earth's surface day in and day out.

This is not to say that all places on Earth would have the same temperature if the axis were perpendicular to the ecliptic. Rather, at any given place the temperature would not vary seasonally because of changes in the length of day or angle of sunlight. The equatorial lands would still be hottest and the polar areas coldest.

Under these conditions, the small changes in heating that occur as the earth changes distance from the sun would be more pronounced than they are today. The distance-related variations in weather on such a world would, however, be much less significant than the changes in seasonal weather we experience.

URANIA

Ninety-Degree Obliquity—Maximal Seasons

The world with the most extreme temperatures throughout the year would be one whose rotation axis lay in the plane of its orbit around the sun (ninety-degree obliquity). This is the world you would make by holding your pen horizontal and fixed in direction as it orbits your head. We will consider this world in the rest of this chapter. Modeling it after the planet Uranus, we will call it Urania because Uranus's rotation axis lies within ten degrees of its orbital plane.

Urania is to be identical to Earth in all respects except that its rotation axis will lie in the plane of the ecliptic. Since the planet has such a large angular momentum, its north and south poles point in fixed directions throughout the year. For concreteness, we set Urania's North Pole to point at the bright star Regulus in the constellation Leo Major. Regulus is the brightest star you see directly below the pot of the Big Dipper. It lays on the ecliptic although, of course, Earth's North Pole never points at it.

Urania has same mass as the earth. The locations of its oceans and continents are identical to those of Earth. A year on Urania would be just as long as a year here; a sidereal day would be twenty-three hours and fifty-six minutes long, just as it is on Earth. (Recall that a sidereal day is measured by the rising and setting of the stars rather than of the sun.)

Urania would also have a moon, identical to our own, located at the same distance as ours is from the earth. However, Urania's moon could be in one of two distinct orbits— either in the plane of the ecliptic or around Urania's equator—and still be identical to our moon. In the latter case the moon would have an orbit exactly perpendicular to the ecliptic.

This issue does not arise for the earth. Because the earth's obliquity is relatively small (twenty-three and a half degrees), our moon orbits nearly in the plane of the ecliptic *and* nearly over the earth's equator. Since laying Urania's rotation axis on the ecliptic puts the plane of its equator perpendicular to the ecliptic, we must make a conscious choice of where to orbit Urania's moon. For guidance we will turn again to the planet Uranus, whose moons orbit in the plane of its equator. We will therefore put the moon in orbit above Urania's equator. This is also consistent with the formation of the moon splashing off that planet, as it is believed to have splashed off the earth.

The orientation of Urania's axis makes the light and climate cycles there very different from those on Earth. Since each cycle is quite complex, we will discuss each separately, beginning with Urania's light and times.

Sunrise, Sunset
Both the daily and the annual motions of the sun would be more complicated on Urania than they are on Earth. Let us compare how the days progress throughout a year on Urania and on Earth as experienced in the Atlanta, Georgia, of both worlds. Atlanta is chosen arbitrarily; except for the exact dates of various events, the same results apply everywhere except at Urania's poles and equator. We begin as spring gently awakens life in Atlanta.

Just as on Earth, on Urania the beginning of spring in Atlanta would find the sun over the equator. The vernal equinox would bring twelve hours of daylight and twelve hours of darkness. During the month centered on the equinox, the moon and sun would have roughly the same positions *relative to each other* as seen from Urania as they have as seen from Earth. Therefore, the tides on Urania created by both the moon and sun during this time would be the same as they are during March and April on Earth. There would be high tides and low tides, neap tides and spring tides.

Whereas during the months following this equinox on Earth the sun rises only to twenty-three and a half degrees north latitude, on Urania the sun would continue to rise higher and higher over the Northern Hemisphere *every day*.

This is because Urania's North Pole would not just tilt generally toward the sun up until the summer solstice, but it would point directly at the sun on that day. Using your pen once more to represent Urania's axis, hold it horizontal in front of your face with both ends equidistant from your eyes. Move it horizontally around your head so that its nonwriting end is now closer to you than its tip. Again be sure to keep the pen pointed in the same direction all the way around. One quarter of the way around your head you will find the nonwriting end pointing straight at you.

In detail, this three-month period as seen from Uranian Atlanta would pass as follows. The first days following the vernal equinox would be similar to the same days on Earth, with the sun rising higher in the northern sky each day. On April 25, if you happen to be outdoors at noontime in Atlanta, you would see the sun directly overhead. This never happens on Earth's Atlanta. For a month before and a month after April 25, Atlanta and other places on Urania at thirty-four degrees north latitude would feel the scorching heat from the sun that on Earth is reserved for the tropics.

From April 25 through May 18 the sun would pass north of Atlanta at noon. On May 18, it would rise and stay up in the sky north of Atlanta without setting for the next sixty-eight days. During this period there would be no nighttime in Atlanta or anyplace north of thirty-four degrees.

It might help you to visualize this situation by noting that today there are constellations, such as Ursa Minor and Ursa Major, that are so near the North Pole star Polaris that they are up all night long as viewed from most places in the Northern Hemisphere. These are called circumpolar constellations. Indeed, if we could see these constellations during the day, they would be "up" then, too. Circumpolar constellations never pass below the horizon. To anyone standing on Earth's North Pole, all the constellations in the sky are circumpolar; from there all the stars move parallel to the ground, neither rising nor setting. To anyone standing on the earth's equator, none of the stars are circumpolar; they all rise along the eastern horizon and set along the western horizon.

On Urania the sun crosses in front of the North Pole star Regulus. For several months on either side of that day the sun would never set as seen from much of the Northern Hemisphere; it would be circumpolar. On Urania the summer solstice would find the sun directly over the North Pole, with Regulus directly behind it. As seen from Atlanta, the sun would be thirty-four degrees above the northern horizon on that day.

The sun would start south beginning the next day as Urania travels farther around the sun, carrying the North Pole away from pointing directly at the sun. The normal day and night cycle in Atlanta would resume on July 25. The sun would pass directly overhead again on August 18, bringing Atlanta another bout of tropical weather. After that, the sun would rise progressively farther to the south.

On the autumnal equinox, September 22, the sun would again be over Urania's equator. The sun and moon would then be related as they are today, the moon would go through its normal phases, and the tides would be similar to those we experience on Earth. (More about the phases of the moon later.) The sun would continue rising farther south every day for the next three months, not stopping and heading northward after reaching twenty-three and a half degrees south latitude as it does on Earth. From the autumnal equinox through the rest of September and into early October, the days would shorten, the sun would rise farther to the south, and the weather would become crisper. By the middle of October, Atlanta would be as cold as southern Canada in winter on Earth.

On November 17 the sun would skim along the southern horizon as seen from Uranian Atlanta. It would be up for less than an hour. On November 18 the sun would be so far south as to be below Atlanta's southern horizon; it would fail to rise at all. By this time Urania would have gone around the sun far enough since the autumnal equinox so that it would not be visible at all from anyplace north of Atlanta for the next several months. Urania's Atlanta would be plunged into a sixty-eight-day period of continuous darkness, softened only by moonlight.

On January 26 the sun would finally peek above Uranian Atlanta's southern horizon again for a few minutes. Periods of daylight would become longer and longer thereafter. The sun would return to the vernal equinox on March 22, and the cycle would begin again.

Lands of the Midnight Sun

There are places on Earth that experience prolonged darkness and prolonged daylight. These are the regions north of the Antarctic Circle (at sixty-six and a half degrees north latitude) and south of the Arctic Circle (at sixty-six and a half degrees south latitude). We call them the lands of the "midnight sun." In contrast, *everyplace* on Urania except around its equator would have extended periods of darkness and extended periods of daylight. As we have just seen, these times would be separated by weeks or months of what we, living on Earth in temperate climes, consider to be normal, twenty-four-hour cycles of day and night.

How long each part of Urania's complex day-night cycle lasts would vary with latitude. As seen from Urania's equator, the sun would never quite stay up or down for a full day; every day would bring some light and some dark. The farther from the equator one traveled, the longer would be the periods of continuous light and dark. At the poles the sun would be "up" continuously for six months and then "down" continuously for the other six.

SEASONS IN URANIA'S MIDLATITUDES

All the major cities and virtually all the human inhabitants of Earth live where there is a daily cycle of light and dark throughout the year. Except for cities built right on Urania's equator, no place on that world would undergo that daily cycle every day of the year. As a result, the seasons on Urania would be far more severe than they are on Earth. Consider the temperature profile of Atlanta on Urania.

Rising higher and higher in the sky, the spring sun would melt ice frozen in Georgia during the long, cold winter. During this period the sun would shine daily as it does on the equatorial reaches of the earth. Rising higher in the sky as the spring wears on, the sun would heat Atlanta's air to tropical temperatures of over one hundred degrees Fahrenheit.

When the sun moves into continuous daylight, it would be north of Atlanta and lower in the sky than it was during the spring when it passes directly overhead each day. Therefore, even though the sun is up continuously during the late spring and early summer months, it would actually provide less heat to Atlanta than it did during the spring and will again in the fall.

When the sun moves southward back over the midlatitudes of Urania during the fall, the land would already be quite warm from the spring and summer heating. Because the sun would once again be directly overhead, Atlanta would again be exposed to searing heat. Like Atlanta, the rest of the midlatitudes, including North America, Europe, and Asia, would be extremely arid then until the sun got low enough in the southern sky to allow rainwater to remain on the ground.

During the time of the autumnal equinox in Uranian Atlanta the weather would be temperate and similar to what we experience in our Atlanta. Shortly thereafter, the sun would descend so low in the sky that the lands of the Northern Hemisphere would begin freezing. By the time the sun sets below the southern horizon, winter would already be brutally cold, since the periods of daylight would have been very short for a month or more. All parts of the planet experiencing dark months would cool until rivers and lakes froze solid. The temperature would get down to over a hundred degrees below zero Fahrenheit near the poles and would stay down there until the sun once again climbed high enough in the sky to start the melting process that would herald the beginning of spring.

The time relationships among cities at different longitudes *and* different latitudes would be different on Urania. On Earth the different time relationships vary only with longitude, not with latitude. The earth is divided into time zones running mostly north-south, with east-west zigs and zags determined by political boundaries. For example, the United States is divided into four time zones. Most of the dividing lines between zones snake around the edges of states, which represent political boundaries.

The biological clocks for virtually all people in a given time zone on Earth are entrained by daylight. The same is not true for people at different latitudes or on different hemispheres of Urania. Therefore, on Urania the logistics of relationships among people in different cities would be much more complex than they are on Earth.

Seasons at the Poles

The seasons at Urania's poles would range to the greatest extremes experienced anywhere on that planet or on the earth. Recall that there are presently ice caps over both poles of the earth. A massive iceberg floats in the Arctic Ocean in the north, and a glacier covers Antarctica in the south. Both of these caps are permanent. Earth's ice caps have built up over billions of years (when Antarctica was farther north, the southern ice cap also floated in the sea).

Cooling at the poles is extreme on Earth because the sun is completely below the polar horizon continuously for six months each year, causing perpetual night. Earth's polar regions receive no direct heat from the sun during this time. Even though the sun is up continuously for six months as seen from each pole, the ice caps are permanent because the sun is never high enough in the sky to melt them. This shows the importance of heat supplied by intense sunlight beating straight down. Because of the earth's twenty-three-and-a-half-degree obliquity, the highest the sun ever rises above the horizon as seen from a pole is only twenty-three and a half degrees.

For six months at a time the sun would also be below the horizon of each pole on Urania. During this time a substantial amount of ice and snow would form there. However, during the other six months, when the sun is "up" continuously, the heating of Urania's poles would be inconceivably strong, very different than it is here. On the summer solstice the sun would be directly above Urania's North Pole. It would remain overhead for several weeks. The sun's heat would pound down mercilessly, melting whatever arctic ice had formed the previous winter. The same would occur in Antarctica six months later. Therefore, the ice caps on Urania would only be seasonal.

The water that is now locked up in the earth's southern ice cap on Antarctica would be in Urania's oceans, thereby raising sea levels by about 125 feet over what they are on Earth. (Since the arctic ice cap is already part of the oceans, it would not help raise sea level when melted.) All Urania's continents would therefore be smaller than Earth's, with shorelines farther "inland." (By the way, Atlanta, Georgia, is 1,050 feet above sea level on Earth. In Urania it would still be on dry land.)

Seasons at the Equator

The equatorial lands would enjoy the most moderate climate of all places on Urania. Unlike on Earth, we have seen that the sun would not be high in Urania's equatorial sky every day. The climate at the equator would be tropical for only about four months out of the year—during the two months centered around each equinox. During the rest of the year the sun would be lower in the sky throughout the day, thereby providing less heat than during the two tropical seasons.

The equatorial lands would also be the only places on Urania where the sun would rise and set every day. The equators of both the earth and Urania have the distinction of being the only places on either planet where one could see all the stars in the sky at one time or another during the year. At all other latitudes, some stars would be permanently below the horizon. In that respect these are just the opposite of the stars in the circumpolar constellations.

At both Earth's and Urania's equator there are twelve hours of daylight and twelve hours or darkness every day. However, at each of the solstices, the sun is over one of Urania's poles. Therefore it would only be a few degrees above the horizon as seen from the equator. During the seasons surrounding the solstices, the sun would provide very little heat and relatively little light to Urania's equator compared with solstice times on Earth.

Since the sun would be up for at least a little while every day, the equatorial regions of Urania would receive at least some heat every day, keeping them warmer than other parts of that planet. It is around the equator, then, that life would probably first emerge onto Urania's surface; there the greatest diversity of life would always exist. We will explore the life on Urania later in this chapter.

Lunar Phases

The cycle of lunar phases seen from Earth occurs because the moon orbits the earth in essentially the same plane as the earth orbits the sun. As we discussed in chapter 2, the lunar phases repeat continuously as the moon slips from between the earth and the sun to behind the earth and back. The phases of Urania's moon would be completely different. Recall that Urania's rotation axis lies in the plane of the ecliptic, while the moon orbits above the planet's equator. As a result, the moon's orbit is exactly perpendicular to Urania's ecliptic, instead of being (almost) parallel to it, as our moon's orbit is.

To follow a cycle of phases for Urania's moon, we begin again at the vernal equinox when the sun is heading north over Urania's equator. On this particular year we place the moon directly between the earth and the sun. The moon is therefore in the "new" phase. A solar eclipse would also occur as a result of the sun, moon, and earth being exactly aligned.

Eclipses would be exceptionally rare on Urania. Because it takes some twenty-eight days to orbit that planet, the moon would rarely be between it and the sun at the time of an equinox. Throughout the rest of the year the sun, moon, and Urania would never be in a straight line, which is necessary for eclipses to occur. Solar and lunar eclipses would each occur only once every forty years or so.

For a few days after the eclipse, the moon would develop into a crescent. It would continue to fill out to a first quarter and then become gibbous (more than half full) before the sun is too far north of Urania's equator. So far the cycle is similar to that of the earth's moon. One subtle change becomes apparent during this time. Since our moon orbits close to the earth's ecliptic, the boundary between light and dark on the moon, called the terminator, always runs north-south as seen from Earth. As Urania's sun moves northward, the terminator on its moon would slowly tilt from north-south to east-west.

When the sun nears the solstice over Urania's North Pole, the side of the moon it illuminates would be centered around the moon's north pole. During this time the sun and moon would be at right angles to each other as seen from anywhere on Urania. Therefore, during the summer solstice one half of the moon would continually be visible from the planet's surface; the moon would not be going through phases because it would never be moving in front of or behind Urania relative to the sun.

The moon would remain half full (a "quarter" moon) for perhaps a month as the sun rounded the North Pole and began heading south. Then the moon would begin to either wax or wane, depending on whether it was on the opposite side of Urania from the sun or on the same side. When the sun next arrived over Urania's equator (autumnal equinox), the moon, as seen from Urania, rather than being new as it was at the previous equinox, would be a waning crescent with about 6 percent of its surface lit. It would never reach the new phase because it would never pass directly between the sun and Urania. Rather, it would next become a waxing crescent, growing to quarter and gibbous before moving back to quarter as the sun headed south toward Urania's pole. The moon would again remain a quarter moon for a month when the sun reaches the winter solstice. This time the southern half of the moon would be lit as seen from Urania. When the sun arrives at the vernal equinox again, the moon would be a crescent showing about 12 percent of its lit surface to Urania.

The crucial point is that Urania would not have a repeatable *monthly* cycle of lunar phases. In fact, even its *annual* cycle of phases as described would be variable from year to year. Every year the cycle of lunar phases would be different, and changing phases resembling those we are used to on Earth would occur only in abbreviated form around the times of the equinoxes.

Tides

Because the moon and sun would be the two major causes of tides on Urania, as they are on Earth, the cycle of tides would also be different there than it is here. Generally, the tides would be lower on Urania than they are here because very high tides occur only when the moon, sun, and earth are in a straight line (new or full moon). We have just seen that new and full moons would be very rare on Urania. During most of the year the moon and sun would be at right angles to each other. At those times the tides would have their smallest range throughout the day, just as they do during similar times on Earth. Only twice a year, around the times of the equinoxes, would tides on Urania be nearly as high as they are each month on Earth. And they would reach peak heights only on those years when the moon happens to be new or full.

LIFE ON URANIA

Since Urania is identical to the earth in all respects save the tilt of its axis, the question of whether life would evolve there hinges upon whether the sun would provide it with enough energy. We will see in chapter 5 that life required ultraviolet radition to begin forming. Once constituted, it has needed heat to keep from freezing to death. It seems likely that in those places on Urania where there would be several months of continuous darkness each year, the early evolution of life would be difficult. Since there would be very little ultraviolet radiation or heat reaching those regions of the planet's surface, any early life that did form during the summers would die during the winters. While this problem also existed around the polar regions of Earth, in Urania it would extend all the way down through the midlatitudes, including where most of North America, Europe, Asia, and Australia are today.

While Urania's equator would not be immune to cold weather during the solstices, it would at least receive daily doses of ultraviolet radiation in its formative millennia before the planet's ozone layer develops. If life begins forming at all, it would be in the waters around Urania's equator. Urania's equatorial oceans, while cooler than those on Earth, would still be temperate and so would provide a suitable environment in which life could develop.

For similar reasons, life would probably emerge onto Urania's relatively mild equatorial shores. As we have just seen, for a few months each year equatorial low temperatures would reach down below the freezing point of water, while equatorial high temperatures would probably get up to around 110 degrees Fahrenheit. Although early surface life on Earth did not have to adapt to such an extreme range of temperatures, life could adjust to it if necessary. Indeed, the life on Earth that migrated from the hot equatorial climes to the midlatitudes have done so.

As on Earth, once life is established on Urania's equatorial landmasses, it would almost certainly disperse over the entire planet, evolving to cope with the extremes of light and dark, hot and cold. The annual extremes of climate in all but the equatorial lands would require the evolution of more hearty plants and animals than those on Earth. Life would eventually even flourish on the polar lands, which, after all, receive tropical-strength sunlight for several months each year.

Vegetation

The midlatitudes and polar reaches of Urania present impressive challenges to all kinds of life. Consider the difficulties faced by trees there. The bright, hot months with the sun directly overhead would be ideal for the growth of tropical plant life. But those same latitudes would experience continuously dark, bitterly cold months that would put enormous stress on such vegetation. On Earth, faced with potentially lethal cold weather, many trees shed their leaves in the fall. The needles of evergreens lose heat and moisture more slowly than do broad, flat leaves and can use sunlight to manufacture food during the winter months, enabling the tree to withstand the cold.

However, needles like those of the evergreen would be more of a liability than an asset on Urania: During the months of darkness the needles would gather no light and still give off heat and moisture. During this time the trees would receive no benefit from the needles, while losing precious fluids through them. Therefore, the long periods of darkness and cold on Urania do suggest that virtually all trees, except perhaps those at the equator, would drop their leaves before winter sets in.

For trees designed to withstand tropical heat the months of subzero weather would be detrimental even if they did drop their leaves. The cells in these trees would be much more susceptible to damage by severe cold than the cells in trees designed especially to withstand low temperatures. To ensure that delicate living cells in the wood stay warm enough to revive in the spring, nature might find it profitable on Urania to provide trees with an alternative heat-gathering system.

One way this biologically expensive, but potentially life-saving, scheme might evolve is as a labyrinth of deep roots that tap into the relatively warm earth below the frost line found in the soil at depths of three to four feet. (The frost line is the level below which the ground does not freeze.) The roots would draw heat up into the trunk of the tree. Such a system is analogous to a central heating system for a skyscraper, in which the heat is carried upward by a network of insulated pipes. Warmed by the ground's thermal energy, the sap in trees and perennial plants on Urania could carry heat upward, thereby providing enough warmth to the living tissue to take the "edge" off winter.

Because of Urania's extended winters, the frost line would be deeper there than it is on the earth. However, the roots of Urania's trees could burrow farther down during the long, hot months with the sun high overhead.

During the hot times on Urania, trees could be kept cool in a similar manner. On Earth the ground below the frost line does not warm much above fifty degrees Fahrenheit, and the same would be true on Urania. During the sizzling summer months the sap could carry down some of the heat that the leaves could not shed directly into the air and deposit it into the soil. Such systems of temperature control would enable plants to live over a wider range of Urania's latitudes.

This use of the root system for heat control would require more fluid in Urania's plants than in Earth's vegetation. The sap would also have to be less sticky on Urania than it is here so that it could quickly transport heat through the tree. Urania's tree sap would more closely resemble blood in its viscosity and speed than does sap on Earth.

This abundant sap might well contain combustible chemicals that, when refined, could serve as a source of fuel when technological societies develop. On Earth we see hints of this property in some saps, such as pine sap. Sap trees on Urania would be far more valuable alive than as dead wood for burning or building houses. From our perspective in a technological society, active and chemically sophisticated trees would be an excellent, virtually unlimited source of renewable energy. Their sap could save the trees of Urania from mass destruction by humans.

Animal Survival During Winter
The mixing of daily and seasonal cycles of light and dark on Urania would profoundly affect the evolution and

behavior of both animal life and plant life. Even more than plants, animals would have to adapt to the change from daily cycles of light and dark to seasonal cycles of light and dark and the accompanying changes from mild to extreme temperatures. We will consider two specific effects of these cycles on animals. First, we will explore how animals might respond to the large temperature changes associated with the sun being "down" for months. Second, after discussing the spread of humans, we will examine the effects on biological clocks of changing from daily to seasonal lighting.

On Urania as on Earth, animals living in the midlatitudes would have essentially three options for surviving winter. They could stay awake and active, they could migrate to warmer climes, or they could hibernate. Staying awake and active during winter when the ground is frozen and most plant life dormant would entail storing food, fattening up, and finding other sources of food and water. On Earth many kinds of animals, ranging from small mammals to deer, moose, and their kin, do this. Landlocked fish do not migrate, of course, nor do some birds. For active animals on Urania, hot spots on the planet's surface such as those in Yellowstone National Park would be thermal oases during the interminable, bone-chilling dark times in much the same way that desert springs on Earth are oases amidst the dryness around them.

Indeed, Urania's months-long dark periods, with their icy cold and thick blankets of snow, would be much harder on animals living far from the equator. In particular, access to water would be extremely limited. The lack of even minimal heating from a weak winter sun, such as occurs on Earth, would cause the ground, streams, rivers, and lakes to freeze even more deeply than they do here. On Urania it's likely that provided with sufficient fat, more animals would hibernate through the long, dark winters than do animals on Earth.

The types of animals that hibernate on Earth run a broad gamut, as does the depth of their torpor. While skunks, racoons, badgers, some bats, and many rodents go through a deep hibernation, bears are lighter sleepers, and some birds such as nighthawks, whippoorwills, hummingbirds, and swifts, some bats, and some other animals sleep even more lightly.

On Earth migrating animals expend large amounts of energy traveling to greener pastures nearer the earth's equator during fall months. Many species of birds, butterflies, mammals, and other creatures make such journeys, leaving the sparse winter food supplies from their summer ranges to the relatively few animals that remain behind.

Migration, a major alternative to hibernation, would require longer treks on Urania than it does on Earth, since more of Urania's lands would be frozen. On Earth migrating animals often travel thousands of miles across several continents to get to warmer feeding grounds each winter. On Urania, with darkness over the entire northern or southern hemisphere, migrating animals would have to travel to the equator or beyond to find suitable food supplies.

This suggests that on Urania island continents such as Australia would not support land animals who need to migrate to equatorial climes to survive the dark periods there. Even as recently as 4 million years ago, North America and South America were not connected. Migration of land animals from North America to the equatorial regions of South America would not have been possible back then, limiting the types of animals that historically lived on Urania's North America.

The longer migratory treks necessary on Urania each year would take more time if animals there travel at the same rates they do on Earth. On Urania, however, the cold weather would descend more quickly upon the midlatitudes than it does on Earth because the sun there would settle more quickly southward (as seen from Urania's Northern Hemisphere). The lower sun would provide less heat than it does on equivalent days on Earth. To survive the rapid onset of winter, migratory animals would have to move faster on Urania than they do on Earth. On Earth, animals use speed primarily to help them hunt and escape from predators, whereas on Urania migration rates would affect evolutionary success. Those animals that migrate fastest would have the advantage of more food at the end of their journey as well as more time to breed and raise young.

In any event, the importance of migration would change the entire relationship between animals and the land from what it is today. Indeed, on Urania the building and fencing off of the Panama Canal would mean annihilation for every species in North America that would need to get to South America for the winter. To be viable on Urania, the Panama and Suez canals would have to be spanned by innumerable migratory bridges.

Spread of Humans from Equatorial Urania

As with other animals, humans on Urania would probably evolve from equatorial ancestors. The same happened on Earth, where our ancestors seem to have come from equatorial Africa. Our forebears were able to spread toward the midlatitudes in small part because Earth's day-night cycle is absolutely reliable.

On Earth evolution provided humans with the flexibility to withstand significant seasonal changes in weather. This ability, combined with the development of clothing and the use of fire, enabled our ancestors to spread all over the planet (except to Antarctica). However, these vicissitudes were nothing compared to the inhospitable climates with which early people on Urania would have to cope.

Staying warm during the long, dark, cold winters on preindustrial Urania would require large amounts of fuel for fires. Until the value of burning tree sap became widely known, the demand for heat would undoubtedly lead to the burning of large quantities of wood. This would create a great drain on the forests of Urania, even greater than that which occurred on Earth, where centuries ago many forests were clear-cut for fuel and never replanted.

Until they learned conservation techniques, early people venturing into Urania's midlatitudes would have to travel as much in search of wood to burn as in search of food. The demand for heat from trees would be so great on Urania that it wouldn't be surprising to see people develop forest-management techniques millennia before they did on Earth. Growing trees for fuel (first as burnt wood and later as sources of sap) would be as important a farming industry as growing food there, since the needs for warmth and light would be so vital to surviving the long dark periods each year.

Housing in the midlatitudes would also be a major problem for early people on Urania. Just as they were on Earth, caves would be the easiest living quarters in which to keep warm in winter and cool in summer. Warmed by the heat from inside Urania, caves would not get as cold as the planet's surface does during the dark winter season. Similarly, during the scorching summers caves would be cooled by the rock surrounding them. It seems reasonable to expect that people would be cave dwellers on Urania for even longer than they were on Earth; with wood for heating at a premium, using caves for homes would make even more sense on Urania than it did on Earth.

Searching for food would also be different on Urania. The locations of the great transcontinental animal migration routes would affect where early people would live, as would the location of water. Early towns and cities would spring up near where the vast herds forded rivers.

When technology permitted it, ocean fishing would

become a vital source of food on Urania because it could be done throughout the year; although lakes would freeze, the oceans would not. Since it would yield a steady supply of food, fishing would be even more important in supporting early people on Urania than it was on Earth.

Large-scale agriculture would face severe challenges on Urania. With good water management, equatorial farming can be done nearly year-round on Earth. Equatorial farming on Urania would be limited to three-month periods twice a year when the sun is high in the sky. Between these times the sun would be too low in the sky to provide sufficient light and heat for most crops to grow. Snow would fall at Urania's equator even during solstice seasons.

While equatorial farms would have two relatively short growing seasons, the midlatitude farms would have one very long season. In fact, on Urania the amount of time available for farming in the midlatitude regions of America, Europe, and Asia would actually be longer than it is on Earth. Farmers throughout the midlatitudes would have six or more months of continuously warm weather during which to farm. Their farming season would include many of the days with "normal" cycles of light and dark as well as the days when the sun remains continuously up over the northern sky. During this latter time the sun would be fairly low in the northern skies of the midlatitudes, so the sun's heat would not be very intense. However, the constant sunlight during this latter time of the year would make up for otherwise cool days.

Animal Migration and the Human Population Explosion

Large-scale farming on Urania would compete for land with large-scale animal migration. Vast herds of animals would eat or trample any fields of planted foodstuffs in their way, so fencing would be essential to protect the agricultural products. But too many fences blocking north-south migration routes would lead to the extinction of innumerable species of animals on Urania.

The growth of human population centers on Urania, with the accompanying need for space and boundaries, would also threaten the survival of migrating animals. One plausible solution would be wide migration corridors for animals that follow the sun. These regions would have to be devoid of farming. But given our experience on Earth, it seems unlikely that people on Urania would stay entirely out of the enormous migrational territories needed by the vast herds making their semiannual pilgrimages. People living in these regions would have to build their homes and cities surrounded by sturdy walls or moats to channel the migrating animals around them. Such boundaries would probably make cities and isolated homes in these areas look like double-prowed boats oriented north and south to force the herds around the inhabited areas.

Synchronizing Biological Clocks

The distinction between animals active during the day (diurnal species) and those active at night (nocturnal species) would become blurred on Urania, since everyplace except the equatorial regions would shift from daily to seasonal lighting. Without being able to change what entrains its biological clocks, life on Urania would be thrown into chaos as the seasons passed. For example, nocturnal animals going to sleep on the day the sun rises for its continuous period of daylight would not have darkness in which to function for months. The converse would apply to purely diurnal animals. Whether migrating or staying in one region, animals that remain active the year round would need to adapt to the change between daily cycles of light and dark and seasonal periods of continuous light or dark. The key to their survival would lie in flexible biological clocks.

The interplay between daily and seasonal light cycles on Urania would wreak havoc with biological clocks designed for use on Earth. As we saw in chapter 1, biological clocks regulate our lives and those of other animals and plants. Implicit in that earlier discussion was the assumption that the clocks are reset or entrained by the sun's light *every day*. Since that condition of daily light would apply only during part of the year on Urania, life cycles would need to be driven by other forces during the months of darkness.

One of the crucial effects of entraining biological clocks on Earth is that all animals are synchronized with one another, regardless of their biological clocks' natural rates. In that way different species react to one another in ways that optimize their chances of survival. The loss of daily sunlight during the dark months on Urania would influence the behaviors of animals as individuals, their relationships with animals of the same species, and their prey-predator connections with animals of different species.

As mentioned in chapter 1, other external influences besides the day-night cycle can entrain biological clocks. However, most of these others involve changes in some factor that is linked to the day-night cycle. Secondary entrainers on Earth include changes in temperature, changes in humidity, changes in sound levels, and changes in air pressure, all of which are affected by the daily presence of the sun.

The one external influence that would change daily throughout the year on Urania regardless of whether the sun rises and sets is the presence of the moon. Just as on Earth, it would rise and set daily. Although the moon's brightness would not be the same each day, the time interval between moonrises would be. Animals living away from Urania's equator would evolve greater sensitivity to low light levels, such as the presence or absence of moonlight, than animals have on Earth. Therefore, moonlight could serve as the entrainer for biological clocks during Urania's prolonged nights.

The fact that moonlight on Urania would not be constant as the moon goes through phases would not pose a problem for its ability to entrain. Studies of animals on Earth show that the light intensity necessary to entrain them need not vary to the same peak each day. It just needs to be cyclical in changing intensities. Lunar days are twenty-four hours and fifty minutes long. This would be the entrainment cycle (or day) during the prolonged nights on Urania.

The moon would not be a suitable entrainer, however, for periods of prolonged daylight. Moonlight is so weak compared to sunlight that during the periods of continuous daylight on Urania the presence or absence of the moon would not be noticeable. Another entrainment mechanism would be necessary for the months during which the sun doesn't set. Note that even when the sun is up continuously, it is not always in the same place in the sky throughout each day. As Urania spins on its axis, the circumpolar sun would sometimes be higher in the sky and sometimes be lower. The sun would go through twenty-four-hour cycles of height in the sky, and this would be as close as the sun would get to changing from day to night in the season of continuous light.

The sun would spread most heat when it is highest in each daily "height" cycle. This is exactly what occurs during different parts of the day on Earth: Near sunrise and sunset the sun's light is coldest, while at noon the sun imparts the most heat. Plausibly, then, the amount of heat deposited by Urania's sun in its "daily" height cycle could serve to entrain biological clocks.

Unsynchronized Biological Clocks

The need to use secondary entrainers such as the moon or changing levels of heat raises an interesting "what if" question about Urania: What if none of the secondary entrainers was suitable for synchronizing life during the prolonged seasons of light and dark on Urania? Recalling that biological clocks tick at different rates in different species, could diverse life forms survive if different species all began living at their natural biological clock rates during parts of each year? This situation would completely change the prey-predator relationships from those that would exist when the sun rises and sets daily on Urania.

The problems associated with unsynchronized biological clocks run in two directions. First, a species that is prey for another could be driven by its internal clock to building homes or mating when its predator is out hunting for it. The prey would be especially vulnerable when its mind is distracted from its foe. Second, species that would normally be predators could enter sleep cycles when species that are normally their prey or who normally are not competitive with them are awake. Then the sleeping predators would become prey. In either case, the food chain would be seriously disrupted when the day-night cycle gives way to long periods of dark or light. This suggests that nature would absolutely need to find a secondary entrainer to keep animals functioning with their biological clocks working in lockstep.

Mechanical Clocks

Early people on Urania would encounter the same dilemma that other animals face in dealing with prolonged darkness or light. The coming and going of the moon or relatively subtle changes in heat that could be detected by other animals would easily be missed by people living in caves or other protected habitats. Living in protective shelters, people not adapted to the midlatitude changes in light cycles would be at great risk of getting out of sync with the rest of the natural world as well as with other members of their tribes and with other tribes.

Without alternative natural entrainers, people on Urania would have an urgent need for mechanical alarm clocks to help them regulate their activities as soon as they ventured to lands far from the equator. The alarm clocks would have to run at different rates at different times of the year. We have argued that during the dark season creatures would be entrained by moonlight. To match animal activity during that time of the year, the people on Urania would need their mechanical clocks to run with lunar-day cycles of twenty-four hours and fifty minutes.

On the other hand, the length of "day" during the season of continuous daylight would depend on the motion of the sun across the sky. This cycle would be twenty-four hours long (just like our day), and so mechanical clocks would have to run in twenty-four-hour cycles during that time of the year.

WHAT IF THE SUN WERE MORE MASSIVE?
GRANSTAR

OUR SOLAR SYSTEM OF THE SUN, PLANETS, AND ENTOURAGE OF moons, asteroids, and comets is 25 trillion miles from the next star, Proxima Centauri. Traveling at fifty thousand miles an hour, it would take a spaceship from Earth 57,000 years to get to Proxima Centauri. Such a spaceship would encounter nothing of substance along the way other than the occasional interstellar iceberg. Outside the ship, the temperature of interstellar space would be nearly minus 450 degrees Fahrenheit. Clearly, the sun is a marvelous thermal oasis for the earth in an otherwise frigid and profoundly hostile universe.

In the first four chapters of this book we have taken the sun's energy emission and its permanence for granted, as we all do in everyday life. This complacency prevents us from developing a true appreciation for just how special our sun is as the source of life-sustaining energy. Although the sun is completely ordinary as stars go, we will see in this chapter that it is absolutely extraordinary when it comes to supporting life on Earth. We will do this by having our earth orbit a star with different properties.

Perhaps surprisingly, many people consider the sun to be something other than a star. Indeed, it does appear bigger and brighter than the pinpoints of light we normally call stars. This leads some people to conclude that the sun is something unique. In fact, the sun is just one of over 200 billion stars in the Milky Way galaxy.

As we will see shortly, a star's mass (total number of particles) is primarily responsible for determining the range and intensity of electromagnetic radiation it emits. These emissions, in turn, determine the star's suitability for supporting life-bearing planets. Over 99 percent of the stars in the Milky Way have different masses from the sun's.* Choosing an alternate "sun," therefore, amounts to selecting a star with a different mass from that of our present sun. In this chapter we will see what life would be like on an Earth orbiting a more massive sun.

How massive a star should we choose? Stars have masses ranging up to one hundred times the sun's mass, hereafter called a solar mass. Not all of these more massive stars would be suitable for supporting the earth. To make a reasonable choice we need to know how stars work and what special properties our alternative sun must have if it is to allow life to form on a planet orbiting it.

ANATOMY OF A STAR

A Star and Its Planets Are Born

Stars have been forming for over 10 billion years in our Milky Way galaxy. They derive from vast interstellar clouds

---

*The only other characteristics of a star that influence how much electromagnetic radiation it emits are its chemical composition and whether it has a companion star locked in close orbit around it. In this chapter we consider only stars that are isolated from their neighbors. Also, since the effects of chemical composition are minor compared to those we will discuss, we will ignore them.

of gas and dust filling trillions of cubic miles of space. Like the one from which our sun originated, these clouds are typically composed of 73 percent hydrogen, 24 percent helium, and traces of the ninety other naturally occurring elements. These latter elements exist in the clouds as dust particles, similar in size to those you often see in a room lit by streams of bright sunlight.

From time to time interstellar clouds collide with one another or encounter shock waves from nearby exploding stars. The clouds are compressed by these interactions until small regions in them become so dense that they begin collapsing under the force of their own gravity. It is inside these cloudlets that stars form.

If the gas and dust are not rotating, it all falls straight to the center of the cloudlet, creating a single, nonrotating star. No planets are created under these conditions. Planets can form only when a collapsing cloudlet is spinning. In that case the outer parts of the cloudlet spin more and more rapidly as they fall inward. (Analogously, an ice skater spins faster and faster as she brings her arms in closer to her body. In a cloudlet, gravity serves the same role as muscle, while the gas and dust are the analog to the skater's rapidly spinning arms.) The outer part of a rotating cloudlet never falls onto the star forming at its center. This matter has too much energy of rotation, and so it settles into a disk in orbit around the young star. This disk then develops clumps forming either another star or a solar system of planets, moons, and other debris. Our interest here lies in a single star and its solar system.

Fusion: A Star's Heartbeat

A star begins forming in the center of a collapsing cloudlet as infalling gas and dust pile upon each other. At this stage it is called a protostar. A protostar's core is compressed and heated as more and more of the cloudlet's mass plummets onto it. Eventually the core pressure reaches about 40 billion pounds per square inch (almost 3 billion times the pressure our bodies feel from the earth's atmosphere pressing down on us). The temperature in the core at this time is around 27 million degrees Fahrenheit. This pressure and temperature are great enough to force the hydrogen atoms in the protostar's core to begin fusing together to create helium.

Normal hydrogen atoms consist of two elementary particles: a proton and an electron orbiting around it. The proton has a positive electric charge, while the electron has a negative charge. It is their opposite electric charges that keeps the electron in orbit. While both particles have exactly the same amount of charge, the electron is considered to be in orbit because the proton has nearly two thousand times more mass than the electron. For hydrogen, the proton is also called the nucleus of the atom.

The internal pressure in the protostar fuses two protons together, creating three particles: a deuterium nucleus, a positron, and a photon. Deuterium is a bound combination of a proton and a neutron (a neutral particle). A positron is identical to an electron except that it has a positive electric charge. Another fusion reaction in the star fuses the newly formed deuterium with another proton to create helium 3 and another photon. The helium 3 atom, containing two protons and a neutron in its nucleus, fuses with another helium 3 atom to create the ubiquitous helium 4, or ordinary helium, with two protons and two neutrons. While all aspects of these and the other fusion reactions in stars are interesting and insightful for nuclear physicists, the really crucial thing about them for the life of the star and for life on planets around the star is the photons they create.

Photons are particles of electromagnetic energy. We will discuss them in more detail later, but suffice it to say here that when created in sufficient number in a protostar's core, they slam into nearby particles with sufficient force to stop the core from compressing any further. Some photons from the core move upward, striking particles above the core and stopping the outer layers of the young star from collapsing further inward. At this point there is a balance between the inward force on the star from its own gravity and the outward force from the fusion-generated photons, called photon pressure.

When the protostar stops collapsing, it becomes a Main Sequence star. This name derives from the fact that such stars form a sequence with increasing mass; the higher-mass stars on the Main Sequence are consistently hotter and brighter than the lower-mass stars. The sun arrived on the Main Sequence 4.6 billion years ago. All stars on the Main Sequence fuse hydrogen into helium in their cores until all that hydrogen is consumed. The rate at which the hydrogen core fuses into helium depends primarily on the star's mass.

The photons created by fusion have energy but no mass. Photons represent the conversion of some of the mass of hydrogen atoms into energy in the fusion process. This means that the newly formed helium atoms have less total mass than the individual hydrogen atoms from which they are made. The amount of mass converted into energy (i.e., photons) in the fusion process is determined by Einstein's famous $E = mc^2$.

This equation says that the energy of amount E released in a nuclear-fusion event comes from the mass of the particles that were fused. The lost mass is denoted m. The mass lost by the particles and the energy of the emitted photons are related by a universal constant, c^2, which is the speed of light multiplied by itself.

The energy created in a star's core eventually leaks out into space, lost forever to the star. This energy loss means that stars lose mass throughout their lifetimes. Our sun is no exception. Over the past 4.6 billion years the sun has lost 0.3 percent of its mass due to fusion. That corresponds to losing the equivalent of 1,000 trillion, trillion pounds.

Whether life could develop on planets around a certain star is primarily determined by how intensely that star emits various kinds of photons (electromagnetic energies) into space. Since knowing about the different types of photons and their effects on life will help us choose a more massive star, we will now examine the photon emissions from Main Sequence stars with different masses.

Photons: Massless Particle/Waves of Electromagnetic Radiation

Photons, the massless morsels of electromagnetic energy created by fusion, are extraordinary and unique particles. Sometimes they behave like massive particles with the ability to transfer momentum to other particles, just like billiard balls. A good example of this is when photons stop the collapse of protostars. Sometimes photons behave like water waves rather than billiard balls. Imagine standing in the ocean as a wave passes around your body, reconnecting after it goes by. Individual photons can do the same thing. A single photon sent toward two adjacent holes will pass through both of them and recombine on the other side! Although they are finite in size, photons do undulate like ocean waves.

Photons can be visualized as tiny, discrete packets composed of waves of energy. The waves inside a photon diminish in height at the leading and trailing edges of packets, which is why the photons are discrete, rather than continuous like ocean waves. All the waves in one photon have the same wavelength, which is the distance from the crest of one wave to the crest of the next.

Regardless of their wavelengths, all photons travel at the same speed. Called the speed of light, it is 186,000 miles per second. Of particular importance to their various roles in forming and nurturing life, photons with different wavelengths have different energies. The shorter a photon's wavelength, the more energy it packs, and vice versa. Since this relationship between photon energies and wavelengths is universal, a photon is discussed in terms of its energy and its wavelength interchangeably. The photons with which we are all most familiar have wavelengths between 16 millionths and 28 millionths of an inch; they are called visible light.

All photons are classified by their wavelengths: Photons with wavelengths longer than $\frac{1}{250}$ of an inch are called radio waves. These are the lowest-energy photons. Photons between this size and 28 millionths of an inch are called *infrared* because they have wavelengths just longer than the red part of the visible spectrum. Then comes the *visible* or *optical* part of the spectrum.

Our brains interpret visible photons with different wavelengths as different colors. From the longest to the shortest visible wavelengths, we interpret the photons as red, orange, yellow, green, blue, and purple (or violet), respectively. These are also the colors of the rainbow, which is just sunlight separated into its basic, or spectral, colors. Each spectral color corresponds to a range of wavelengths, blending smoothly from one to the next. All the other colors we see are created by combining different intensities of the spectral colors.

The photons with the next shortest wavelengths are *ultraviolet,* with wavelengths between 16 millionths of an inch and 400 billionths of an inch. Then come the *X rays,* with wavelengths down to 400 trillionths of an inch. All shorter-wavelength photons are called *gamma rays.* Gamma-ray photons pack the greatest energies. The photons created in a star's core by fusion are virtually all gamma rays, yet most stars emit more visible light than any other kind of electromagnetic radiation.

Star "Light"

Clearly, newly formed gamma rays must undergo a transformation between the time they are created in a star's core and the time they reach its surface and charge out into space. From our point of view this is just as well; gamma rays are absolutely deadly to life as we know it. If the sun emitted mostly gamma rays, there would be no possibility for life on Earth. This would apply to any planet around any other star as well. How, then, do gamma rays get converted to visible light and other radiations compatible with life?

The answer is buried in the observation that as soon as each gamma ray is created, it bangs into a nearby atom in the star's core. The photon is absorbed and then re-emitted by the atom, after which it soon strikes another atom, which absorbs and re-emits it. So begins a long, arduous journey from the star's core up to its surface.

Photons lose energy to the atom they strike during each collision. Part of the photon's energy thus lost serves to push the atoms outward, thereby counterbalancing the inward force of gravity. As mentioned earlier, this is what keeps the star from collapsing. Photons therefore emerge from atoms they hit with less energy than they had when entering them. *As the gamma rays lose energy in their travels toward the star's surface, they transform into lower-energy photons.*

As different photons travel upward, they encounter different numbers of atoms and, therefore, have different numbers of collisions. The photons reaching a star's surface have therefore lost different amounts of energy. Different parts of the electromagnetic spectrum are represented differently at the star's surface. Relatively few photons leave the star as radio waves. More of them go as infrared radiation, and most photons emerge as visible light. Many are ultraviolet, while relatively few are X rays and even fewer are gamma rays.

In other words, photons reaching each star's surface have a wide range of energies throughout the entire electro-magnetic spectrum. This process of leaving a star's core and rising to the surface takes time. For example, photons leaving the sun's surface today were created in its core a million years ago.

Wien's Displacement and the Colors of Stars

The amount of each type of electromagnetic radiation that each star emits depends on the star's mass. This distribu-tion of different types of photons is called the star's spec-trum. As suggested earlier, the radiation from every star has a peak of intensity. Depending on the star's mass, the peak is in the infrared, the visible, or the ultraviolet part of the spectrum.

Most stars have peak emission in the visible part of the spectrum. These stars, while emitting all colors of the rain-bow, are associated with the color at which they emit the most photons. This color is brighter than any other visible color, and so the star appears to give off only that one color. For example, since the sun appears yellow, it is natural to assume that it emits only yellow photons. However, the sun actually gives off all colors of photons.* The point is, of course, that colors other than yellow are emitted less

---

*Actually, the sun's intensity peaks in the blue-green part of the spectrum. However, the earth's atmosphere scatters sunlight differen-tially. That is, the shorter-wavelength purple, blue, and green pho-tons do not all pass through the atmosphere. More of these photons are scattered by the air than any other colors, and so they do not reach our eyes from the sun. The effect of not seeing all the green, blue, and purple photons is to change the apparent color of the sun to yellow. So as not to confuse matters any more than necessary, we won't use this fact in what follows, but will continue referring to all objects having the same temperature as our present sun as appearing yellow.

intensely by the sun and are therefore overwhelmed in our eyes by the sun's yellow photons.

Exactly where a star's peak of radiation lies depends on the temperature of its surface. Our sun, appearing yellow, has a surface temperature of ten thousand degrees Fahrenheit (5,800 degrees Kelvin). Every star and indeed every object that has a surface temperature of ten thousand degrees Fahrenheit has the peak of its electromagnetic radiation in the yellow part of the visible spectrum.

In 1893 Wilhelm Wien discovered a relationship between the surface temperature of objects and their color peak. He found that cooler objects have peaks in the infrared, hotter objects peak in the visible part of the spectrum, and the hottest objects peak in the ultraviolet realm. This relationship between surface temperature and spectral peak of intensity is now called Wien's displacement law. It shows that the peak of brightness moves from infrared for the coolest stars through red, orange, yellow, white, blue, violet, and ultraviolet for hotter and hotter stars.

Although white is not a color of the rainbow (it is how our brains interpret light of equal intensities from all visible colors), it falls in the middle of this rainbow scheme. This happens because stars that ideally would appear green give off essentially equal amounts of all visible colors, making them appear white instead. Wien also showed that hotter objects are intrinsically brighter than cooler objects of the same size.

Stars and Paper Clips

It is worth emphasizing that stars are not the only objects that obey Wien's displacement law. In fact, almost everything you can think of behaves this way. Consider, for example, a paper clip. You normally see the clip by the light it *reflects*. This reflected light is irrelevant for now and we hereafter ignore it. We are only interested here in the electromagnetic radiation that the paper clip emits from energy stored within it.

The paper clip does not have fusion occurring to make the photons it emits; it gets and stores energy from the sun and other sources. Of course, if you look at the clip in a completely darkened room, you won't see it glowing. That is because our eyes are too insensitive to see the few visible-light photons the paper clip emits when it is that cool. At room temperature the peak of intensity for photons is in the infrared part of the spectrum. If our eyes were sensitive to infrared photons, the paper clip would be quite visible even in a dark room. The numbers of gamma-ray, X-ray, ultraviolet, visible-light, and radio-wave photons emitted by the clip at room temperature each second are minuscule compared to the number of infrared photons it emits.

Now heat the clip briefly. The first color it appears to glow is red as its peak of electromagnetic radiation shifts out of the infrared and into the regime of visible light. Since red photons have the lowest energies of visible light, they are associated most strongly with the coolest objects that have peaks in the visible part of the spectrum. The red light overwhelms the light of different colors the paper clip is also emitting. Heat it a little more and the clip appears orange and brighter. Further heating makes it appear yellow and brighter still. Continue to heat it and it becomes white, then blue, and finally purple (assuming it doesn't melt), getting brighter all the time. This is exactly how a sequence of stars of the same size but with different surface temperatures would appear.

The Brightness of Stars

More massive Main Sequence stars are brighter than less massive ones for two reasons. First, as we have just seen, the surfaces of more massive stars are hotter, and therefore each part of a hotter star's surface emits more of all types of photons than the same area of a cooler star's surface. Second, more massive stars are always physically larger than less massive stars. Therefore the more massive ones have more surface area from which to radiate more photons.

More massive stars are hotter because they create more energy each second through higher fusion rates in their cores than do lower-mass stars. Greater fusion occurs because more massive stars have more gravity, which compresses them harder and creates higher pressures in their cores. The higher pressure causes more fusion than occurs in lower-mass stars. The greater the fusion rate, the larger the number of photons being created. Since all these photons eventually reach the star's surface, the more massive stars emit more photons into space each second, helping to make them brighter than lower-mass stars.

THE NEW SUN'S MAXIMUM POSSIBLE MASS
While the most massive Main Sequence stars have one hundred solar masses, a star this large could not support life on Earth. There are two factors that limit the maximum mass a life-supporting star could have to below one-hundred solar masses. First is the effect of a star's electromagnetic radiation on the process of planet formation. Second is whether the star will last long enough for life to evolve on a planet orbiting it. Since it is not a priori obvious which of the two effects is more restrictive, we will briefly examine both.

Planet Formation or Ultraviolet Expulsion
The earth formed more than 10 million years after our present sun joined the Main Sequence. Its formation was delayed because the collapse of the solar system–forming cloudlet was not uniform. The cloudlet's central regions formed the sun much more rapidly than its outer regions formed the disk from which the planets were created.

The problem that faces planet formation around much

more massive stars is that during the 10-million-year interval between the star's formation and that of its planets, the star would rid its neighborhood of the disk gas and dust. Just as photons inside the star push against atoms (to prevent the star from collapsing in on itself), photons outside the star can push nearby atoms of gas and dust in the disk outward. Sufficiently intense photon emissions literally blow the surrounding disk of gas and dust out of orbit before it has time to clump and form planets.

Of all photons, ultraviolet are the most effective in pushing matter out of orbit around young, massive stars. Astronomers calculate that the ultraviolet radiation emitted by stars with more than ten solar masses is so intense that virtually all the dust and gas around them is ejected before planets can form. Therefore, the more massive sun we seek must be a star with less than ten solar masses.

Stellar Lifetimes

The other possible limit on the mass of life-supporting stars we need to examine is derived from how long it takes each star to consume all the hydrogen in its core. Although more massive stars have much more hydrogen fuel in their cores than do lower-mass stars, their stronger gravitational compression makes them fuse it into helium at higher rates than do lower-mass stars. And so, higher-mass stars complete the Main Sequence stage of fusing all their core's hydrogen into helium more rapidly than do lower-mass stars. As noted earlier, the greater fusion rate is part of the reason why higher-mass stars glow more brightly and more hotly than do lower-mass stars.

When any Main Sequence star converts all of its core's hydrogen into helium, fusion within it ceases. Without new photons to provide the pressure that keeps it stable, such a star's core begins to collapse. As it collapses, the hydrogen just outside the core follows inward and is compressed enough to begin fusing into helium. This new shell of fusing hydrogen outside the core creates new photons that, being closer to the star's surface, provide even greater outward pressure than did the photons created in the core. (In other words, each collision between one of these shell photons and a nearby atom is more effective in pushing that atom outward than it was for photons created in the core.)

Instead of the entire star collapsing inward, as one might expect, the powerful and copious shell photons actually push the outermost layers of post-Main Sequence stars even farther out than they were originally. Amazingly, stars actually expand while their cores shrink. As the outer layers swell, this gas moves farther from the shell fusion and, like a camper moving away from a fire, cools down. As the surface of the star cools, it becomes redder. Such a bloated, post-Main Sequence star is called a red giant.

Planets orbiting a star at the distance the earth is from the sun are burned to cinders during the star's red giant phase. Indeed, calculations show that when the sun enters this expansion phase in about 5 billion years, its surface will move outward 200,000 times farther than it is today. This will bring the surface of the sun out to the vicinity of the earth.

How long must a star remain on the Main Sequence before intelligent life could evolve on a planet orbiting it? The earth is our only source of data on this question. It took 4 billion years before the earth was capable of supporting life on land. It took over 600 million years more before life had evolved sufficient complexity to be aware of its own existence. We, of course, are the present heirs of that long process.

Considering the time it took to make the air breathable, to evolve land creatures, to allow for recovery from mass destruction of species by asteroid impacts and diseases, it seems that 4.6 billion years is a plausible minimum life for a planet before "people" evolve on it. Therefore, the earth's new star would have to contain sufficiently *low* mass to keep it on the Main Sequence at least that long. Calculations show that the most massive star still on the Main Sequence after 4.6 billion years has only one and a half solar masses.

We choose this to be the mass of our new sun. It may seem at first that adding only 50 percent more mass to the sun is a disappointingly small change. But we will soon see that even this alteration leads to major changes in the earth's relationship to the sun.

The More Massive Sun's Appearance

The major differences between our sun and the more massive one would be in their colors, temperatures, and brightnesses. These changes are directly associated with the increased rate of fusion that accompanies greater stellar mass. The new sun's surface temperature would be 12,000 degrees Fahrenheit. Appearing blue-white, it would be five times brighter than our present sun. This increased brightness would occur in part because the more massive sun would have 70 percent more surface area than our sun and in part because each part of the surface would emit more energy than the corresponding surface of our sun.

We begin by creating a planet identical to the earth, with an identical moon, at the same distance from the new sun as we are from our sun. The first question we will address is whether the earth, so formed, would be able to support life.

EARTH AT ITS PRESENT DISTANCE FROM A MORE MASSIVE SUN

Runaway Greenhouse Effect

There is no good substitute for vast quantities of water as a nursery for complex life. As we saw in chapter 1, copious supplies of water are needed to serve as the solvent in which life evolves from raw chemicals. A star's major function in making a planet habitable is to keep the planet's surface temperature in the range where water can remain liquid. Since under virtually all conditions some water will either evaporate or condense to a solid, the sun must actually enable water to exist in all three of its phases: solid, liquid, and gas. This allows the water to cycle from the oceans to the air to the land and back again to the oceans.

The first problem we encounter on the earth orbiting the hotter, more massive sun is an increase in the surface temperature of the planet. This change would greatly affect the water cycle just mentioned. The earth's average surface temperature today is forty degrees Fahrenheit. If the earth were instantly transported to orbit the hotter, more massive sun, the average temperature here would immediately rise to sixty-five degrees Fahrenheit. For the earth *born* in orbit around the more massive sun, the temperature would be higher still, and the surface of the planet would be uninhabitable.

As described in chapter 1, oceans began developing on Earth only after the planet cooled sufficiently to enable rainwater to condense on the surface without immediately being re-evaporated. We saw that our earth was hot because of the impacts it received from planetesimals, because of radioactive elements on its surface, and because of heat leaking out from inside it. The sun was never mentioned there in the context of overheating the earth. With the end of impacts and the sinking of the radioactive elements toward the earth's center, our planet cooled, carbon dioxide and water escaped from the rocks and into the atmosphere, and eventually oceans rained out of the sky.

However, water vapor and carbon dioxide are greenhouse gases. The brighter, more massive sun would send more energy through the earth's early atmosphere than our sun did. The atmosphere, being rich in these greenhouse gases, would retain more heat than it did from our cooler sun. As a result, the vapor in the atmosphere would be kept so hot that it would not be able to condense into water on the surface. Such a greenhouse effect would be self-perpetuating; once established, it would keep the water in vapor form forever. The surface of the earth at our present distance from the more massive sun would be bone-dry while surrounded by oceans of water vapor in the air. This is called the runaway greenhouse effect; it is occurring today on Venus.

The increase in the earth's temperature to sixty-five degrees does not take into account the runaway greenhouse effect. Including the greenhouse heating, the average temperature of the air would be well over two hundred degrees Fahrenheit. Since the oceans would never form under such circumstances, life would never evolve to transform the earth's atmosphere into the nitrogen-oxygen one we have now. The earth orbiting the more massive sun at our present distance would be forever enshrouded in a dense carbon dioxide–dominated atmosphere with permanent cloud cover like our sister planet, Venus.

THE EARTH FARTHER FROM A ONE-AND-A-HALF-SOLAR-MASS SUN

The New Location

As it is today, our earth just barely avoided this scenario. To enable a planet around the more massive sun to be cool enough to have oceans, we need to move it farther out in space. The ideal location for a life-supporting planet around the new sun is three and a half times farther away than we are. This new orbit would be located between Mars and Jupiter, out in the present Asteroid Belt.

Living among asteroids would pose grave problems, however. The planet would pull asteroids onto its surface much more frequently than our earth attracts the infrequent debris passing by today. Even occasional impacts have devastating effects on life, as seen by the extinction of the dinosaurs and other mass extinctions after such collisions on our planet. In the presence of hundreds of thousands of asteroids bombarding the more distant earth over billions of years, it is most unlikely that life there would be able to evolve as far as it has on Earth.

This problem is actually a red herring; the Asteroid Belt would not exist in the solar system with the more distant earth in its midst. The asteroids formed in a region of the present solar system where the planetesimals did not have enough mass to form a planet. (Contrary to popular opinion, the asteroids are not the debris of a destroyed planet. Rather, they are the debris that lacked enough gravity to ever consolidate and form a planet.) If the earth condensed where the asteroids are today, the asteroid material would have become part of the earth during its formation, and the Asteroid Belt as such would never have existed.

This more distant Earth is our focus for the rest of this chapter. We call it Granstar to emphasize the increased mass of the sun it orbits.

The New Year

A year on Granstar would be longer than a year on Earth in part because Granstar is farther from the sun and in part because the new sun is more massive than our sun. The speed at which a planet orbits a star is determined by the star's mass and the planet's distance from the star. These two effects compete with each other.

Two planets that are the same distance from stars of different mass orbit with different speeds; the planet around the more massive star completes a circuit (a year) more rapidly than the planet around the less massive star. Also, two planets at different distances from the same star complete their orbits at different rates. That is, the two planets have years of different length; the farther planet always takes longer to go around than the closer planet. Granstar's greater distance to its sun would have more effect on the length of the year there than would its sun's greater mass. Granstar's year would end up being five and a third of our present years long.

Since Granstar's axis would have the same tilt as Earth's axis, Granstar would go through the same seasons as does the earth. However, the longer year would mean that each season would be five and a third times longer than it is on Earth. Hidden in these numbers is a potential barrier to the evolution of life on Granstar: runaway glaciation.

Runaway Glaciation

The earth has glaciers and other permanent ice and snow cover in those places where the temperature is too low to ever melt the frozen water. Almost all of the sunlight striking these white regions of the earth is reflected right back into space rather than being absorbed by the planet and thereby helping to heat it. Where the sunlight strikes water or darker parts of the ground, more heat is absorbed by the planet. On earth, when the snow cover melts in the springtime, the ground again absorbs energy from the sun. The sun's heat allows the frigid earth to thaw out and carry on supporting life.

The problem for Granstar comes from the possibility of reflecting too much sunlight during the prolonged winters, thereby starting a cycle of excessive cooling. The longer winters on Granstar would enable more snow to fall, making more of that planet whiter than our earth becomes during wintertime. As a result, Granstar would reflect more sunlight, absorb less heat, and therefore be colder overall than the earth. This would lower the temperature more than occurs during our relatively short winters. The cooler Granstar becomes, the more snow and ice would fall on it.

This cycle of cooling—snowfall—more cooling—more snowfall can be unstable. Like the Ice 9 of Kurt Vonnegut's *Cat's Cradle,* once such a cooling spiral gets started, it might not stop. This situation is called runaway glaciation. If it occurred there, more of Granstar would become covered with thick glaciers and icebergs during the long winters. The vast stretches of white glacial ice on Granstar would reflect more sunlight in the spring than the earth does. Therefore, the sun would be less able to thaw Granstar's enormous expanses of ice and snow compared with the area on Earth thawed each spring. In that case much more of Granstar would remain permanently blanketed in ice and snow than covers Earth.

Granstar would actually need a higher average temperature than our earth in order to prevent excessive ice and snow cover from occurring each winter and initiating runaway glaciation. Like our earth, Granstar would be precariously balanced between runaway glaciation and the runaway greenhouse effect.

LIFE ON GRANSTAR

Even with Granstar positioned where its water would remain liquid, life there would face more significant hurdles before it could blossom than did life on Earth. These problems would occur because Granstar's massive sun would emit more of all kinds of radiation than does our sun, as well as more of the high-energy particles called solar wind.

Ultraviolet Impact on Early Life

Granstar's more massive sun would emit almost one hundred times more visible light and heat than our sun does, as well as many thousands of times more ultraviolet radiation. The increase in emissions of the various types of electromagnetic radiations is not uniform when going from a cooler star to a warmer one.

Moving Granstar farther out in the solar system would enable that planet to keep its temperature in the right range. Unfortunately, even out there it would be bathed by ultraviolet radiation over a thousand times more intense than the ultraviolet radiation that strikes the earth today. This increase in ultraviolet radiation would affect every stage of evolution on Granstar.

As briefly mentioned in chapter 1, the effects of ultraviolet radiation on life begin in the very earliest stages of evolution. The earth's early atmospheres, unlike today's, allowed ultraviolet radiation to reach the planet's surface. Along with lightning, this radiation provided much of the energy necessary for the basic chemical combination of atoms and inorganic molecules that eventually led to life on Earth. The blending of molecules began in the earth's atmosphere, with the resulting larger molecules raining into the oceans where the process continued.

But as far as life is concerned, ultraviolet radiation is a double-edged sword. While that radiation supplied the energy that enabled organic molecules to form in the first place, it also supplied the energy that broke up these molecules into simpler ones; ultraviolet radiation destroys its own creations. Organic molecules that did not fall into Earth's oceans quickly enough were destroyed by continued ultraviolet radiation. Once in the oceans, early complex molecules were screened from ultraviolet radiation by water, which absorbs ultraviolet photons. Hovering between the dangerous ultraviolet on the surface and the colder, energy-deprived depths, life in Earth's early oceans developed within some thirty feet of the water's surface.

Whereas the ultraviolet radiation striking the earth's oceans is absorbed by the first few feet of water, the heavier ultraviolet bombardment on early Granstar would penetrate several tens of feet deep before being completely absorbed. The initial formation of life in Granstar's atmosphere would begin as it did on Earth. But then the stages in the oceans would have to take place deeper down to protect the nascent life from destruction by the more intense ultraviolet radiation. This would be possible in part because deeper parts of Granstar's early oceans would be warmer than they were on Earth. However, the pressure in the deeper water would affect the shapes and biologies of the creatures evolving there and, consequently, those that emerge onto land later in Granstar's life.

In chapter 1 we discussed how life moved onto Earth's landmasses only after the atmosphere was converted from being dominated by carbon dioxide to being dominated by nitrogen and breathable oxygen. Omitted from that discussion was how the sun's lethal ultraviolet radiation also had to be screened before surface life could develop here. Without something in the atmosphere to block ultraviolet radiation, early land animals on Earth would have developed lethal illnesses such as cancer. Our ancestors would have died out en masse, and the earth's surface would have remained barren of most types of life that exist here today.

Ozone

The protective screen that blocks out ultraviolet radiation began forming in the earth's upper atmosphere some 2 billion years ago when oxygen began accumulating in the air. Ironically, the sun itself was responsible for creating the barrier to its own lethal radiation. Earth's primary protection against ultraviolet radiation is a form of oxygen called ozone. Oxygen in the air normally exists in bonded pairs called molecular oxygen. The sun's ultraviolet radiation breaks apart some of these molecules into their constituent oxygen atoms. The freed oxygen atoms then combine with other oxygen molecules to form triple oxygen molecules called ozone.

The sun created ozone throughout much of the earth's

upper atmosphere. The ozone layer is thickest some seventeen miles above the surface. Once formed, this ozone began absorbing ultraviolet photons in large numbers, preventing the radiation from getting to the earth's surface. Fortunately for life on Earth, ozone is an exceptional ultraviolet absorber, allowing fewer than one ultraviolet photon in a thousand, trillion, trillion, trillion from getting to the earth's surface. With such an effective screen in place, the danger from ultraviolet radiation to life on land here dropped dramatically and the evolution of surface life proceeded. However, as we are all learning today, the atmospheric ozone layer is extremely precarious.

The problem for maintaining ozone on Granstar (and Earth) is that its chemistry is very complicated. Ozone is continually being created and continually being destroyed by natural chemical reactions as well as by sunlight. Increased ultraviolet radiation from Granstar's more massive sun would certainly create more ozone in its atmosphere than our sun creates around the earth. But the destruction rate for ozone surrounding Granstar would also increase compared to here. Ozone is removed naturally in two ways. First, ultraviolet photons from the sun are absorbed by ozone, thereby breaking it apart. Second, nitrogen and hydrogen (from water vapor) gases combine with ozone and convert it back to molecular oxygen. The question for life on Granstar is: Would the creation process for ozone there be sufficient to offset its destruction, as it is here? Or would the sun's lethal ultraviolet radiation overwhelm whatever ozone does form around Granstar and saturate the planet's surface?

Because our knowledge of ozone dynamics on Earth is far from complete, that question does not yet have a definitive answer. It seems likely, however, that the increased destruction of ozone there would probably prevent that planet's ozone layer (high in the atmosphere) from ever being thick enough to stop all the ultraviolet photons passing through it.

This being the case, more ultraviolet radiation would penetrate to Granstar's surface than penetrates to Earth. On its way down, some of this radiation would change Granstar's atmospheric chemistry, making it different from the chemistry of Earth's lower atmosphere. This radiation would create ozone and other ultraviolet-absorbing compounds near Granstar's surface.

While this added ozone might seem life-protecting, ozone is a highly toxic substance to breathe. It is extremely reactive with other molecules, destroying lung and other tissue. It does so on Earth today, which is why cities with high ozone levels often warn their citizens of the danger. The significant concentrations of ozone *near Granstar's surface* would require evolutionary protection for life both from contact with the gas and from breathing it.

The changes in the lower atmosphere brought about by the more intense ultraviolet flow from Granstar's sun would help diminish the amount of ultraviolet radiation reaching Granstar's surface. Nevertheless, much more ultraviolet radiation would reach Granstar than reaches the earth; that danger would have to be immediately addressed in the creation of surface life there.

Ultraviolet Radiation and Surface Life

Since the ultraviolet radiation on Granstar would be strong during all daylight hours but would drop significantly when the sun goes down, by far the easiest way for surface life to establish itself on Granstar without having to evolve biologically "costly" protection is to do so at night. Nocturnal creatures, buried in sand or sleeping in caves during the day, would be able to move over the land safely after the sun goes down. In all likelihood, then, the first great explosion of animals on Granstar would be nocturnal creatures.

Diurnal life (creatures active during the daytime) could evolve several characteristics to help it avoid the dangers of ultraviolet radiation on Granstar. While cumbersome, inorganic shells would be relatively impervious to ultraviolet radiation. Living under their shells like turtles, some creatures would be safe from the radiation and still able to function during the day. Others might evolve biological "sunscreens" that would permeate the outer layers of their skin. A suitably thick layer of fur would also assist in shielding from the radiation.

While protecting skin from ultraviolet radiation is relatively straightforward, protecting eyes is far more complicated. After all, the coverings over eyes need to be transparent so that visible light can pass through them, while the chemical defenses against ultraviolet damage may not be transparent.

Eyes of animals on Earth today are especially sensitive to ultraviolet radiation. Even the relatively low-intensity ultraviolet radiation that penetrates to the earth's surface causes cataracts and other blinding illnesses. This ultraviolet radiation, which gives us suntans and sunburns, is also the *least* harmful ultraviolet. The more dangerous, higher-energy ultraviolet radiation emitted by the sun is completely blocked by the earth's atmosphere. Nevertheless, eye and skin problems are becoming more common on Earth as the ozone layer is disrupted because of human-made emissions in the atmosphere.

Eyes evolving on Granstar would require much more protection than do eyes on Earth because of the greater intensity of all types of ultraviolet radiation that would reach that planet's surface at all times. Visorlike brows and deeper eye sockets would help minimize the amount of ultraviolet radiation to which eyes are exposed. Thicker eyelids would help keep the radiation away from sensitive eye surfaces. In the best of all possible worlds eyes would evolve a clear membrane with a clear biological sunscreen to absorb the radiation. As this layer of the eye was killed, it would slough off and be continuously replaced by newly grown protection.

In any event, protection of any kind is rarely complete. Animals on Earth, ourselves included, can withstand only limited exposure to hazardous environments. Whatever defenses against ultraviolet radiation did develop on Granstar, they would not be perfect. Ultraviolet radiation is so dangerous that it would be a great asset to Granstar's life to know when the radiation levels were excessively high or when safe exposure times had been exceeded. Creatures on Granstar would probably evolve natural ultraviolet monitors, like the X-ray radiation badges worn by people working in nuclear power plants and by some medical workers. These monitors would warn animals when they need to avoid sunlight while their bodies recover from the ultraviolet doses they have already absorbed.

The Seasons of Life

The longer year on Granstar would greatly amplify seasonal events. We have already noted that many more lakes and rivers would freeze solid during Granstar's winters than do on our earth. But fish on our earth cannot be frozen and revived. So, on Granstar those bodies of water that completely freeze would not be habitats for such fish if they are to live more than one year.

In order for long-lived aquatic life to thrive in Granstar's smaller bodies of fresh water, it would have to evolve the ability to remain frozen alive for long periods (several Earth years). Otherwise, since these bodies of water would freeze solid they would contain only creatures living one season. Since Granstar's year is over five times longer than Earth's, living through one warm season on Granstar amounts to living for several Earth years. This would give most freshwater life plenty of time to mature and reproduce before dying.

Deeper freezing during the winter on Granstar would also force more animals to migrate farther than on our earth, although neither as far nor as fast as they would on Urania. Hibernation on Granstar would be more expensive than on Earth in terms of the energy and fat storage required to keep animals alive during longer winters. Food-storage capabilities there would have to increase dramatically in hibernating animals such as bears, requiring them to be larger in order to store more fat.

Animals active during the winter on Granstar would require larger food caches. The food chain would have to evolve to cope with this need by having trees produce more nuts and other edibles, creating more nutritious nuts, having more trees grow in each area, having wintering animals eat more flexible diets, having fewer wintering animals or species, or, in all likelihood, a combination of these things.

This need for increased vegetation for animal consumption would be fulfilled in part by more growth during Granstar's longer summers. Since both warm and cold seasons would increase in length equally, Granstar would essentially scale up the agricultural activities that presently occur on our earth to fill the longer year.

Life Cycles

Perhaps the most important effect on animal life of Granstar's lengthened seasons would be the change in fertility cycles. The majority of large animal species on Earth have one offspring or one litter a year. This gives the parents time to nurture their young until the children can survive on their own. If animals mature at the same rate on Granstar, with its year over five times longer than ours, then young born in the spring there would become independent of their parents early enough in the year for females to have other offspring the same year. In that case, animals that give birth to only one youngster or litter each year would be at a disadvantage against competitors that have several litters each year. All other things being equal, having more offspring implies that more children will become sexually mature and reproduce.

However, the survival and success of different species goes beyond the relative numbers of children they breed each year. Survival must also take into account the absolute numbers of animals living on the planet at any time and the lengths of their lives. This raises certain questions: Is the maximum length of an animal's life determined by the number of years (that is, seasonal cycles) it lives, by the total number of days it lives, or by some absolute time that is independent of the cycles of the planet? In terms of Granstar, the question becomes: Would animals that evolve there live the same absolute length of time as life does on Earth? Or would they live the same number of *years* as animals do here? In this latter case, animals on Granstar would live five times longer than animals on Earth, in an absolute sense.

If lives are longer on Granstar, then having many more infants born each year would not be as beneficial as it first seems. There are two reasons for this apparent paradox. First, animals that live longer take more time to mature. We frequently see this on Earth, where animals that live ten or twenty years are able to fend for themselves within a year of birth, while animals that live much longer need more nurturing. This is exemplified by humans, who are not set loose from their families until they are teenagers.

On Granstar young animals who need more time to mature would require more attention from their parents. Yearlings would have less chance of becoming adults if their parents had many offspring each year. More children means less food available for each, as well as less protection against predators and less training from the parents. Especially in light of the longer winters, increased competition with their siblings would endanger animals that develop slowly.

Second, even if animals on Granstar develop as quickly as they do here and leave their parents in much less than a Granstar year, increased numbers of babies each year would only be desirable if more of the youngsters normally succumbed to the long winters. Otherwise, Granstar would rapidly become overpopulated. Competition for space and food between species and between members of the same species would be endemic. The "law of the jungle" would be even more violent there than it is on Earth.

The Not-So-Naked Ape

Both the longer seasons and the more hazardous ultraviolet radiation would affect human biology on Granstar. If the long winters lead to more aggressive competition for survival among animal species, the early people on Granstar would also have to be more contentious in order to survive. After all, if the hunted animals on Granstar are more aggressive and cunning than the equivalent animals on Earth, so, too, would the people who hunt them need to be. On the other hand, given their competitive evolutionary background, how successful would people on Granstar be in becoming "civilized"? The meek would have a hard time inheriting Granstar.

As the discussion of ultraviolet radiation suggests, humans on Granstar would need more protection from the sun than our present skins offer. Whereas we humans lost our fur long ago, the intense ultraviolet bath on Granstar would probably prevent that step from ever happening there. Our counterparts on Granstar would almost certainly be covered with protective hair.

Ultraviolet Radiation and Society

We are facing the predicament today of Earth's ozone layer being depleted by emissions from our manufacturing and energy-generating processes. While we will explore this matter in the last chapter of this book, a comparison between the increased ultraviolet radiation now reaching the earth's surface and that reaching Granstar's surface is worth considering.

Just as we have done, people on Granstar would create and use technology that would affect the radiation-shielding ability of the atmosphere. But if the shield were weakened even slightly, the higher ozone flux impinging on Granstar's air would threaten to flood the planet's surface with lethal doses of ultraviolet radiation. It is highly unlikely, then, that civilization on Granstar could allow itself to damage the ozone layer nearly as much as we have done. The damage that the first industrialized people on Granstar would inadvertently cause to their planet's ozone layer would come back to haunt them much more rapidly than the damage we have wrought has returned to haunt us.

We have already argued that Granstar's ozone shield would be more precarious than the earth's. Even small amounts of damage due to industry on Granstar could lead to radiation-caused illness before the people there have developed the scientific and technological abilities to understand and correct the problems they are creating. In that sense, we are lucky. We knew about the existence of ultraviolet radiation at the end of the nineteenth century and about its effects on life early this century, before serious damage to the ozone layer had occurred. Had Earth's ozone layer been severely altered a hundred years ago, many people and animals here would have died from the effects of radiation-induced illnesses even before the cause (ultraviolet radiation), much less the cure (maintaining the ozone layer), had been discovered. When increases in skin cancers and eye problems caused by (then-unknown) ultraviolet radiation began occurring on Granstar, how would society there respond? To what would they attribute the diseases?

In the same way that weather forecasts today often contain information about smog levels, it is likely that weather reports on industrialized Granstar would include information on the daily level of ultraviolet radiation.

Solar Wind

Besides emitting more electromagnetic radiation of all types, the more massive sun would send more particles, called the solar wind, to Granstar. As we discussed in chapter 1, the earth is surrounded by magnetic Van Allen belts. On its way to the earth the solar wind from our sun reaches the Van Allen belts and is deflected and captured by them. As a result, most of these particles do not enter the earth's atmosphere where they could damage the ozone layer and reduce our protection from ultraviolet radiation.

Granstar would have Van Allen belts with the same strength as the earth's. However, even allowing for Granstar's greater distance from its sun, the more intense solar-wind particles would frequently fill Granstar's Van Allen belts to overflowing. The particles would then cascade down into the atmosphere, damaging the ozone layer. Just as on Earth, the excess solar wind entering Granstar's atmosphere would also create auroras (northern and southern lights). In fact, auroras would be a daily occurrence on Granstar.

The Dangers of Space and Air Travel

Ultraviolet radiation is not the only potential radiation problem for people on Granstar, especially when they venture into space. The levels of X rays and gamma rays normally discharged by our sun usually pose little threat to the health of humans journeying into space from Earth today. However, our sun periodically emits sufficiently high levels of radiation to endanger the lives of astronauts in space. Granstar's more massive sun would continually emit such intense levels of X rays and gamma rays that present-day spacecraft would not provide astronauts sufficient protection.

The simplest protection would be to completely line spacecraft with lead, which effectively absorbs X rays and gamma rays. Unfortunately, lead is very dense and therefore so heavy that sufficient shielding would have prevented early rockets from being able to lift useable spacecraft into space. Space travel for people on Granstar would have to wait until more powerful rockets were developed there than were needed in the first decades of space flight for humans here.

The X- and gamma-radiation levels from Granstar's more massive sun would present grave dangers even for passengers and crew in airplanes traveling at just thirty thousand feet, where civilian aircraft regularly cruise today. Indeed, radiation from our present sun penetrating into the atmosphere makes air travel mildly dangerous even on Earth. In light of these radiation dangers from the sun, it is likely that even high-technology civilizations on Granstar would stay on the planet's surface much longer than we did.

A Comet's Tale

The greater radiation and emission of stellar wind from Granstar's sun would have two positive side effects among all their drawbacks. Both relate to comets. First, the massive sun would create more comet tails than does our sun. Second, the comets would provide astronomers with much more information about how stars work than does our sun. While creating comet tails doesn't sound particularly beneficial, we will see that it could be a matter of life and death.

Comets are debris left over from the very earliest stages

of the solar system's formation. They are composed of ice, carbon dioxide (dry ice), frozen methane, and frozen ammonia, all mixed together with rocky and metallic dust, pebbles, and boulders. Typical comets are believed to be irregularly shaped and only a few miles across; astronomers envision them as dirty icebergs in space orbiting the sun. There are millions (perhaps billions) of comets out beyond the orbit of Pluto swarming around the sun like mosquitoes. This ensemble is called the Oort comet cloud, after astronomer Jan Oort, who first proposed its existence in 1950. The Oort cloud may extend halfway to the next star. Indeed, many astronomers expect that most stars have similar icy halos. Comets in the Oort cloud do not receive enough heat from our sun to evaporate and form tails.

Occasionally two comets in the Oort cloud collide, sending one of them toward the inner solar system. Passing inside the orbit of Uranus, our sun's heat starts vaporizing the comet's frigid surface. This creates a spherical atmosphere around the comet, called its coma. Moving even closer to the center of the solar system, the comet encounters the solar wind and more electromagnetic radiation rushing out from the sun. This outflow pushes some of the vaporized gas of the coma outward, creating the tails.

There are often two tails, one of gas and one of dust particles freed from the comet's icy grip. The tails always point away from the sun. Some comets pass close to the sun only once, but others get locked into orbits that bring them back over and over again. Halley's comet is one of the latter type. It passes close to the sun about once every seventy-four years. Each time the sun evaporates some of a comet's surface, the comet shrinks.

Granstar's more massive sun, with its much stronger radiation and solar wind, would create comas (spherical gas halos) around comets farther away than does our sun. Comet tails would also be longer, brighter, and more frequent than they are today. These frequent beacons would add a new dimension to Granstar's night sky.

Eventually comets remaining near the sun are completely vaporized, providing they do not hit a planet or moon first. The positive contribution that Granstar's sun could make to life on that planet is to completely annihilate one or more comets that would otherwise hit the planet. Such impacts by comets are believed to have caused some of the mass extinctions of life on Earth, and they would do the same on Granstar. Preventing even one such event from happening there would considerably advance the speed with which life would evolve. Fewer impacts mean less time that Granstar's plants and animals would have to spend recovering from the shock of a mass destruction.

The second benefit from Granstar's sun, information about how it works as a typical Main Sequence star, comes from observations of the interaction between the solar wind and comet tails. The solar wind does not come out of the sun uniformly. Rather, it comes in waves or spurts, depending on the activities occurring inside the sun and on its surface. For example, the more sunspots there are on the sun's surface, the gustier the solar wind. Sunspots are literally holes created by magnetic fields erupting through the sun's surface. As the flow of solar wind from our sun changes, so too does the way it interacts with comet tails. Fluctuations in the solar wind appear as changes in the shapes of comet tails, similar to the cloud ripples caused by the winds on Earth.

As soon as astronomers on Granstar realized that the changes they would be seeing in comet tails were caused by changes in their sun's activity, they would have a powerful tool for exploring solar physics. For example, astronomers would see a distinct eleven-year cycle during which comet-tail activity rises and falls. This cycle would correlate with weather and auroral activities on Granstar.

Periodic changes in comet-tail sizes and lengths (here or in Granstar's solar system) are caused by the eleven-year sunspot cycle on the sun's surface. The number of sunspots varies from zero to nearly two hundred throughout the cycle. Astronomers would observe that times of longer and more active comet tails would also be associated with times when there are many sunspots. If people on Granstar evolved the natural ultraviolet monitors mentioned earlier, they would also learn that active comet tails occur when the ultraviolet hazard on Granstar is greatest. By combining all these factors, astronomers on Granstar would learn more of the details about their sun's effects on the solar system and on their planet than did our early astronomers. They would even be able to predict from the behavior of the comet tails when ultraviolet radiation was going to be particularly strong.

The More Massive Sun—a Mixed Blessing

The more massive sun, the source of energy allowing life to exist on Granstar, would clearly be seen as a more threatening body than we perceive our sun to be. If nothing else, its ultraviolet emissions would be a constant reminder of the delicate balance of life on Granstar. Therefore, the sun would not be seen as the unsullied fountain of pure radiance it was taken to be by early philosophers and theologians on Earth. A more equivocal sun in the sky, one that demands more attention and creates more fear than ours does, would motivate more accurate explanations of its activities than were needed by early humans on Earth. This information would enable Granstar's people to forecast potentially dangerous changes in the sun so that they could take precautions.

As a result, people might well develop science earlier on Granstar than they did here. The necessity of truly understanding the sun might help humans avoid the equivalent of our last two thousand years, during which religions have suppressed scientific development, especially any facts that made the heavens appear less than perfect.

The More Massive Sun's Old Age

As noted earlier in this chapter, we chose the one-and-a-half-solar-mass sun because it would still be shining 4.6 billion years after it formed. After that, of course, it becomes a red giant. Granstar's sun would not engulf that more distant planet as our sun will the earth. However, Granstar would be overheated and overirradiated when its sun became a red giant. All life on Granstar would surely be killed at that time.

We noted previously that our sun has enough fuel in its core so that this end-of-life crisis will not occur for another 5 billion years. Assuming that humans survive these early millennia of high technology, we will undoubtedly migrate to other star systems billions of years before the earth is in danger from the expanding sun. On the other hand, Granstar's more massive sun would become a red giant star just about when sentient creatures—people—first evolve there.

Imagine how differently we humans would act if understanding of our sun's physics brought the realization that it would start expanding and scorching us any day. This is precisely what the people of Granstar would face. Without a doubt there would be an all-out, worldwide effort to migrate to other stars. The lives of billions of people and the innumerable animals, insects, and plants they would take with them would depend on the abilities of scientists and engineers to design and build huge spaceships to move life off Granstar before it was destroyed by the sun.

Where would they go? As of this writing, we astronomers have not yet identified a single Earth-like planet around any other star. Even when such planets are found, we must determine which, if any of them, have suitable atmospheres and surfaces for human habitation. We would also need to know if our future homes are already occupied by sentient creatures who might resent our intrusion. While we on Earth hope we will have the luxury of being able to decide when and where to go in space, people on Granstar would not be so fortunate. They would need to emigrate to any habitable planet they could find.

WHAT IF A STAR EXPLODED NEAR THE EARTH? ANTAR

IN A.D. 1054 A NEW STAR APPEARED IN THE CONSTELLATION TAU-rus, the Bull. Unlike normal stars, this one increased in brightness and within a week blazed so intensely as to be visible from Earth during the day. After a month of maxi-mum radiance, the visitor began to fade. Two years later it was gone from sight.

It took nine hundred years for astronomers to correctly explain this event. Today we know it was a supernova, the mightiest of all explosions in the universe. Astronomers have located the remnant of that A.D. 1054 explosion. Now called the Crab nebula, it is an expanding cloud of gas and dust some 35,000 trillion miles from Earth. Light from the Crab has been traveling for 5,900 years to reach the earth. For convenience, rather than giving its distance in miles, astronomers say the Crab is 5,900 light-years away from Earth—one-fifth of the distance to the center of the Milky Way galaxy, in which the solar system and all the other visi-ble stars reside.

The Milky Way is a disk-shaped galaxy containing some 200 billion stars as well as interstellar gas and dust. The galaxy also has several spiral arms, but each arm contains so much obscuring gas and dust that astronomers have not been able to see it all or map its farther reaches. Therefore we do not know how many spiral arms the Milky Way contains: There are at least three and probably more.

Besides the emotional trauma that may have accompanied it, nothing particularly devastating happened to the inhabitants of the earth as a result of the 1054 explosion. However, if such an event occurred closer to the earth, the effects could be considerable—even lethal. In this chapter we will explore what would happen if a star did, indeed, explode near the earth. To make sense of the enormity of this event and its aftermath, we must transcend the everyday definitions of *nearby* and *explode* and bring them into the realm of astronomy.

ASTRONOMICAL DISTANCES

When we say two houses are near each other, we most likely think of them standing a few yards apart. Two cities near each other are probably separated by a few tens of miles. Our intuition on astronomical distances is set by the moon and sun. The moon is typically 240,000 miles away in its elliptical orbit around Earth. At that distance, sunlight takes one and a quarter seconds to reach the earth after reflecting off the moon. The sun itself is 93 million miles from Earth; light takes eight and a third minutes to reach us from it directly.

However, when an astronomer says two stars are near each other, the separation involves tens of trillions of miles or more. In fact, our nearest stellar neighbors, in the constellation Centaurus, are over 25 trillion miles away. Light from them takes over four years to get here. If one of those nearest stars exploded, its remnant gas and dust would spread out in a shell that would take 150 years to cross the earth's path. Four light-years is a typical separation between "nearby" stars in our neighborhood of the Milky Way galaxy. The emptiness of space becomes even more apparent when you realize that there are only some twelve thousand stars within a sphere 600 trillion miles (one hundred light-years) in diameter centered on the earth.

Most supernovas that occur in the Milky Way are too far away for us to see them. The problem is not just distance. The plane of the Milky Way galaxy is full of vast interstellar clouds of gas and dust. The light from distant supernovas is absorbed and scattered by these clouds, just as sunlight is absorbed and scattered in clouds in the air. Instead of seeing supernovas from the far reaches of our galaxy, we see the clouds they illuminate. Supernovas are estimated to occur once every thirty years or so in each of the several billion spiral galaxies like our Milky Way. Imagine sitting in a spaceship far above the Milky Way, its majestic spiral arms visible through a porthole. Imagine also that time were sped up so that an hour aboard our spaceship is equivalent to 100 million years in the galaxy below. From that vantage point you would see the Milky Way swirl below like a giant spiral fireworks display with nearly a thousand stars exploding every second all over the spiral arms.

Among the stars you would see exploding from your spaceship are several that are "near" the earth. Within the next hundred million years (which is short as far as astronomical time scales are concerned) Sirius, Canopus, Vega, Rigel, Altair, Spica, and Deneb, among others, will detonate. Indeed, some of these might explode in fewer than a million years, depending on how long they have already existed. There is also another source of stars that are now very far away but that could come close and whose eventual explosions could affect the earth.

The solar system, along with all the other stars in the Milky Way, orbits the center of our galaxy. At least once every 65 million years, the stars pass through one of the galaxy's spiral arms where their paths are altered by the gravitational force from the greater concentration of matter they encounter there. In this way some stars that are presently "far" from us can be sent toward the solar system, while some nearby stars are sent away from it. New stars can also form near us as the solar system is passing through the spiral arms. The stellar explosion whose impact we will explore in this chapter could be one of two kinds: a planetary nebula, which is the term for explosions of stars like the sun, or a supernova. To help us decide which to use and where to position the exploding star, let us consider the power of such explosions.

ASTRONOMICAL EXPLOSIONS

We generally think of explosions in terms of war footage, volcanoes, or the simulated blowups we see in movies and on television. The most horrific explosions for most of us are nuclear bombs, sending mushroom clouds miles into the sky and wiping out life over tens of square miles of the earth's surface. However, such explosions are utterly minuscule compared to the energy emitted by a star during its Main Sequence life, to say nothing of when it explodes.

Consider, for example, the energy normally emitted by our sun. We discussed the sun's light in the last chapter. You may recall that it is energy created a million years ago in the sun's core by the same fusion process of compressing hydrogen into helium that occurs in hydrogen (thermonuclear) bombs. Every second the sun fuses 600 million tons of hydrogen into helium. In one second the sun emits more energy than all the nuclear weapons of all kinds ever built combined. If all the nuclear weapons on Earth were sent to the sun and exploded on its surface,* the explosions would probably not even be visible amidst the other energy emitted there.

When stars explode, and virtually all of them do, they emit even more energy than they do during their lives on the Main Sequence. The type of explosion a star undergoes depends primarily on its mass. Most stars with more than four solar masses explode as supernovas such as the one that created the Crab nebula. Within a few years a typical supernova emits as much energy as our sun gives off during 8 billion years of its Main Sequence life. Here's another way to look at it: A supernova is as powerful as 10 million, billion, billion, billion tons of TNT exploding at once! For a few weeks such an explosion would be brighter than all the hundred billion other stars in a typical galaxy combined. In February 1987 such an explosion was seen in the Large Magellanic cloud, a small galaxy only 160,000 light-years from Earth. Called Supernova 1987A (the first supernova seen in 1987), it temporarily doubled the brightness of that galaxy. By observing this supernova, astronomers were able to confirm many of the basic elements of the theory of supernovas developed over the past forty years.

Stars with fewer than four solar masses explode as planetary nebulas. This is the fate of the sun, which will explode in about 5 billion years. Such an explosion is millions of times less powerful than a supernova; the amount of energy and matter emitted by a planetary nebula is insignificant compared to that given off by a supernova. If we are going to subject the earth to the force of an explosion, we might as well let it face the worst that nature has

---

*The sun has no solid surface. It is entirely filled with hot gas. We speak of the surface as the level of gas that emits the light we normally see. This is called the sun's photosphere.

to offer. So saying, we choose to detonate Antar, a twenty-solar-mass star located fifty light-years from the earth. (Antares, Antar's namesake, is a nineteen-solar-mass star located a more comfortable 325 light-years from Earth. Antares will supernova in the not-too-distant future. Because it is farther from Earth than Antar, the effects of Antares's explosion on Earth will be about one-fortieth those described below.) We consider now why supernovas occur and what happens during them.

THE SUPERNOVA PROCESS

Antar would lead a short, dramatic life. During a mere 8-million-year stint on the Main Sequence, it would convert its entire core from hydrogen into helium. Deprived of new photons to keep it stable, (see chapter 5), the core would begin to collapse. As it did, hydrogen in a shell just outside the core would be compressed enough to begin fusing into helium.

During collapse, the pressure in Antar's core would build until it became great enough for the helium there to fuse into carbon. This new fusion would provide new photons that would temporarily halt the core's contraction. While the helium core burned (i.e., fused) into carbon, it and the shell of burning hydrogen just outside it would provide enough new photons to actually push Antar's outer layers even farther out than when it was on the Main Sequence. Expanding, the star would move into the red giant phase, as discussed in chapter 5.

After a million years, Antar's core would be entirely transformed into carbon. Fusion would then cease and the core would collapse once again. The newly formed helium in the shell surrounding the core would follow it inward. As this helium shell began to fuse into carbon, a new shell of hydrogen just outside it would begin fusing into more helium. Meanwhile the carbon core would compress until carbon fusion began, thereby creating oxygen. This stage would be complete in less than a hundred thousand years, a blink of an astronomical eye.

The next step would be for the oxygen core to collapse, allowing surrounding shells of carbon, helium, and hydrogen to fuse. The oxygen core would be compressed by Antar's outer layers until it began fusing into neon. The neon core would then collapse until it began fusing into magnesium. Another cycle of collapse and fusion would generate silicon. Finally, the core would fuse into iron. As the core converted into these denser and denser elements, the number of shells of fusion outside it would increase. These numerous shells would create enough photons to push the star's outer layers beyond the red giant phase. If Antar were located at the center of our solar system, its surface would be pushed out beyond the orbit of Pluto. Such a star is called a supergiant. Like an onion from hell, our twenty-solar-mass supergiant star would contain shell upon shell of fusing elements surrounding its core.

When Antar's core transforms entirely into iron, the supergiant's future becomes very bleak very quickly. Once again, as fusion stops, the core begins to collapse and heat up. However, iron fuses differently than all the lighter elements; it absorbs photons rather than emitting them. When iron fusion commences, the photons already in the core are *absorbed* by the fusing iron, producing no new ones. This loss of photons cools the core, decreases the outward force it exerts on the rest of the star, and forces the entire star to contract faster and faster. The core is in for the ride of its life.

As the core collapses in on itself, the iron nuclei are literally dissolved by the intense pressure into increasingly lower-mass elements, such as oxygen, carbon, and helium. Aha, you might think, with lower-mass elements again available there, the core could begin fusing and creating photons once more, thereby stopping its own collapse. This does not happen because the collapse of the iron core is so fast and the breaking up of nuclei, called photodisintegration, occurs too rapidly for fusion to take place. The iron core transforms until all that remains of it is hydrogen and helium. After millions of years of painstaking evolution, Antar's gravity would compress its core back to helium and hydrogen in less than one second.

Actually, the star's problems are only just beginning. The collapsing core experiences such great pressure from the layers of mass over it that the core's electrons become fused with the protons in the helium and hydrogen nuclei. Each proton-electron pair is transformed into a neutron. This fusion is accompanied by the emission of a particle called a neutrino. Incredibly, within one second after the iron core begins to collapse, much of the helium and hydrogen core is converted into neutrons. The neutron-rich core continues to contract, converting more and more of the atoms there into neutrons.

The core collapse occurs so rapidly that the onionskin layers outside don't have time to follow inward before the last scene of this explosive drama plays out. In this final step, the particles in the core come to be squeezed so tightly together and at such high temperatures that they literally bounce off one another and rebound violently outward. This rebound creates a tremendous shock wave that slams into the fusing onionskin layers outside the core. Along with an added push on the star's outer layers from the outward-bound neutrinos, all of the onionskin shells burst outward, generating the supernova. (Technically, this is called a Type II supernova. Type I supernovas are described in some of the books on astrophysics listed in the bibliography.)

The onion layers of the star, in which different elements were being formed before the blast, continue fusing during the supernova. While expanding outward, they fuse into existence the most dense elements in the known universe. Indeed, virtually all the matter on Earth and in the rest of the universe (except the hydrogen, some of the helium, and even less of the lithium and beryllium) is created in the fusing onionskin layers of stars before and during supernovas and planetary nebulas. In this sense, we are all truly children of the stars.

Each supernova blast emits incredibly large amounts of electromagnetic radiation, some of which had previously existed in the star's interior and some of which is created in the supernova process itself. Much of this radiation is visible, but all parts of the electromagnetic spectrum are represented, including the highly lethal gamma and X rays. These photons, traveling faster than the expanding shell of matter from the star, become the vanguard of the supernova's effects. Astronomers always detect supernovas by the light they emit, rather than by the particles ejected in the blast. The shells of matter thrown out from the exploding star are collectively called the supernova remnant, of which the Crab nebula is the most famous.

Following the rebound, Antar's core would consist of about four solar masses of neutrons. Along with the energy it would lose upon hitting the onionskin layers, the core would have enough gravitational force to keep itself from expanding outward with the star's outer layers. The core, cooler now, recontracts. But this time it does not bounce back. The core's mass is so great that there is no known force in nature powerful enough to stop it from collapsing in on itself under the force of its own weight.

The core would rapidly contract, getting denser and denser, until it became so compact (a few miles across) that its gravity would be too strong to allow anything to escape from it. It then becomes a black hole, about which we will have more to say in chapter 8. (If we had chosen Antar to have been a Main Sequence star with between four and eight solar masses, the core would have been less massive. It would have recondensed until the neutrons got so close together that their mutual repulsion, called neutron degeneracy pressure, stopped the collapse before it became a black hole. The result of such a collapse is to create a stable, nonfusing remnant called a neutron star.) With the supernova under way, we turn now to its effects on our planet.

GEOLOGY ON EARTH AFTER A NEARBY SUPERNOVA

As noted previously, Antar is located fifty light-years from Earth. This is essentially the minimum distance at which we would want a star to supernova. Considering the titanic forces released in this outburst, it should come as no surprise that the high-energy electromagnetic radiation from such an explosion detonating closer than this would immediately kill virtually all life on Earth.

The people living on Earth would know nothing of Antar's explosion until fifty years after it occurred. It would take that long for the first light from the explosion to cross the void of space and reach the earth. During that interval, Antar would appear as it always had in the sky—as a bright, blue-white supergiant. Then, without warning, the doomed star would grow tremendously luminous, fifty times brighter than the moon and only eight thousand times less bright than the sun. It would be brighter than the light from all the other stars in the night sky combined.

Intuitively you might expect the light from an explosion this powerful to outshine even the sun. It would not, however, because the brightness of different stars depends strongly on their distances from Earth. Imagine two candles, one twice as far away as the other. The farther one would have to emit four times as much light in order to appear as bright as the nearer one. The sun is some 3.25 million times closer to us than the exploding star. The distant supernova would actually have to emit 10 trillion times as much light as the sun to appear equally bright. However, at its brightest, our supernova would emit only between 10 and 100 billion times as much light as the sun. Therefore, the relatively weak sunlight would still outshine the supernova.

Gamma Rays and X Rays

The tremendous flood of deadly X rays and gamma rays accompanying the visible light from the supernova would be the first cause of death to life on Earth. While our atmosphere absorbs all the X rays and gamma rays from the sun, the radiations from the supernova would be so intense that many of these photons would reach the earth's surface. Eventually, this radiation would fall to a level that the earth's atmosphere could once again absorb, ending the most dangerous part of the supernova's impact. We will examine the specific effects of this radiation on life later in this chapter.

Ultraviolet Radiation and the Ozone Layer

The first blast of ultraviolet radiation from the supernova would destroy the earth's ozone layer within a matter of days, transforming the ozone primarily into atomic oxygen. With the removal of this protective barrier, ultraviolet radiation would saturate the earth's surface. The supernova's ultraviolet photons would also be accompanied to the earth's surface by ultraviolet radiation from the sun. This latter energy is normally prevented from reaching the earth by the ozone layer.

As the intensity of ultraviolet radiation from the supernova diminishes, sunlight would begin repairing the ozone layer. In other words, while high levels of ionizing radiation from the supernova would destroy the ozone, lower levels of such radiation help it reform, as we discussed in chapter 5.

As a result, both the ozone layer and the ultraviolet-radiation level would eventually return to normal at the earth's surface, probably over the course of several decades.

The brightness and emission of all the electromagnetic radiation from the supernova would peak after a month. It would then fade, but would be visible for millennia as an expanding cloud of gas and dust in the night sky. Fortunately for life that survives the first onslaught of high-energy radiation from the supernova, the remnant cloud would primarily emit visible light, with more limited emission of the more dangerous ultraviolet, X rays, and gamma rays.

Blast Wave

While the supernova's electromagnetic radiation would get here in only fifty years, the particles blasted up from the star's outer layers would take considerably longer. The bulk of this matter would be ejected into space by the supernova at speeds of ten thousand miles a second (36 million miles an hour), nearly twenty times slower than the emitted photons. Consequently, the supernova remnant would take at least a thousand years more to reach us than did the radiation. In fact, it would take even longer because supernova remnants are slowed down by collisions with interstellar gas and dust they encounter as they spread outward.

Happily, by the time the blast wave of particles from the supernova reaches the solar system, it would have become so thin and diffuse that it would probably do little, if any, damage to the earth or its atmosphere. Recalling that supernovas are where the dense elements in the universe are created, we could expect the supernova remnant to deposit a thin but exotic mix of elements into our upper atmosphere. These would eventually fall to Earth and affect the chemistry of both the ocean and the soil.

Cosmic Rays

Supernovas may be the source of a third type of emission, one that would affect life on the earth. These are extremely high-speed, high-energy particles called cosmic rays. Astronomers believe that at least some cosmic rays originate in supernovas. The name cosmic *rays* is a historical artifact dating back to their discovery in 1912. At that time their true nature as particles was not known. Rather, they appeared to be some sort of radiation. In fact, they are ordinary atomic particles moving at extraordinary speeds.

The majority of cosmic rays are protons, with the rest being electrons and nuclei of helium or other atoms. Cosmic rays are different from the run-of-the-mill gas particles in a supernova remnant because of the tremendously high speeds at which they travel, often over 90 percent of the speed of light. Cosmic rays from the nearby supernova in this chapter would reach the earth five to ten years after the electromagnetic radiation from it first arrives here.

Cosmic-ray energies are much greater than those of any particles normally existing on Earth. For example, a typical cosmic ray has enough energy to light a small light bulb for one second, whereas it takes over 600,000 trillion electrons flowing through wiring in your house to do the same thing.

Since cosmic rays are charged particles, they are affected by any magnetic fields they encounter in their journey through space. In particular, the lowest-energy cosmic rays from the supernova would be captured by Earth's magnetic Van Allen belts. These particles would fill up the Van Allen belts, which would then leak vast quantities of particles down through the earth's atmosphere, creating spectacular auroras.

More energetic cosmic rays would pass through the belts unscathed and plunge into the upper atmosphere. Upon encountering the air, cosmic rays with intermediate energies would slam into nitrogen or oxygen molecules in it, shattering the air molecules into smaller pieces. The debris ejected from the collision would shower downward, striking other air molecules. Some of these would also break up and move downward, creating a cascade of particles that would eventually reach the earth's surface. These cosmic-ray showers are called secondary cosmic rays, while the cosmic rays from the supernova are called primary cosmic rays. Created by primary cosmic rays from other supernovas, from the sun, and from other, as yet unknown sources, these showers happen even today. They are the source of about 10 percent of the environmental radiation to which we are normally exposed. This background radiation would increase dramatically for several years after the supernova's intermediate-energy cosmic rays arrived here.

The highest-energy cosmic rays from the supernova would reach the earth intact, as do the highest-energy cosmic rays from other sources today. These would break up atoms of the objects they strike on the earth's surface.

Remnant Radiation

Finally, the supernova remnant, that shell of gas and dust emitted by the blast, would begin radiating energy from a variety of sources. These include the radioactive decay of some elements created in the supernova, interactions between the remnant and the ambient interstellar gas it encounters, and interactions between the remnant and the magnetic fields that exist throughout the galaxy. These secondary effects from the remnant would be irregular, but ongoing for thousands of years. If nothing else, they would cause the remnant to continue glowing in our night sky.

Life on Earth Immediately After the Supernova

Within days of the arrival of the initial dose of X and gamma radiations from the supernova, individuals in virtually all species of plants and animals would begin dying of radiation poisoning. The same thing occurred in Japan after the bombing of Hiroshima and Nagasaki at the end of World War II and at the Chernobyl nuclear power station meltdown. It is possible, although less likely, that entire species would be immediately annihilated. This radiation would cause many internal cancers over the succeeding years, leading to many more deaths among people and other life forms. At the same time, ultraviolet radiation reaching the earth's surface would cause an astronomical jump in the rates of skin cancer and cataracts. Both the internal and external damage caused by the electromagnetic radiation would be enhanced by the arrival a few years later of the primary and secondary cosmic rays from the supernova.

The damage done to life by both the electromagnetic radiation and cosmic-ray particles would also disrupt global food chains. In the oceans large quantities of plankton and other microscopic organisms at the bottom of the chain would be killed. As a result, many of the larger aquatic species that feed on these smaller organisms would starve. Similarly, much of the surface plant life would wither. The first few crops after the event would be meager, and many herbivorous animals would starve as a result. This would, of course, lead to the death and dislocation of carnivorous and omnivorous animals at the top of the chain.

Mutations
Surviving plants and animals by the millions would undergo genetic mutations due to the supernova's radiation. These alterations would begin with the very next generation conceived after the radiation arrives here. As with mutations that occur in the normal course of evolution, most of the genetic changes would lead to immediate death for the altered plant or animal. As with normal mutations, a

few changes would be beneficial, thus enabling the posses-sors of these new genes to thrive. Although there would be countless deaths due to genetic failure, there would be a relatively large number of successful mutations—relative, that is, to the normal state of evolution.

During "normal" times, species have the opportunity to adjust to one mutation in their food chain at a time. If, for example, a mouse developed a secretion on its fur that was poisonous to all predators, the mouse would be able to increase in numbers relative to its population before the new protective mutation occurred. Rather than being in the middle of a global food chain, the mouse would suddenly be at the top of its own, local chain. Animals relying on the mouse for food would have to find alternative nourishment or face extinction. Equally as important, the poisonous mouse population would rapidly grow and eat up all of its normal food sources. It would then face extinction itself. A new food chain would be established by the time another major mutation occurred to one of the life forms involved.

Unlike the single alteration faced by the food chain to which these new mice belong, postsupernova food chains would be disrupted by new strains of life at all levels simul-taneously. It would be virtually impossible for such changes to be incorporated in existing prey-predator relationships. Rather, these relationships would all break down. Nature would have an enormous task rebuilding the hierarchies of life after the supernova.

Short-Term Survival of Humans

Astronomers would immediately understand the implica-tions of the growing light in the sky. As soon as they spot-ted it, they would recommend that everyone take protective measures against the effects of the high-energy radiation that accompanied the visible light. While everyone would be injured internally by the initial dose of X rays and gamma rays, it would be worthwhile seeking protection from the continuing radiation hazards, especially since the ozone layer would be seriously depleted for years to come.

By far the most effective screen against the radiation would be the earth itself. The rock would absorb virtually all of the ultraviolet rays, X rays, gamma rays, and most of the cosmic rays. The greatest number of people would survive if they all moved underground for a few years. However, with over 5 billion souls living on Earth, this option would clearly be open to only a relative handful of people; neither the room nor the facilities underground exist for extended living by so many people.

The next best options are to sheathe buildings in radiation-absorbing materials such as lead and to stay indoors as much as possible for the next few years. Lead, among the densest elements, is particularly effective at absorbing any radiation that strikes it. From a practical standpoint, lead is unfortunately one of the heaviest elements per volume. A typical American wood-frame house would face perilous structural problems under the weight of a lead coating and lead-impregnated windows.

Engineers would need to develop a lightweight alternative to protect housing and to impregnate clothing so that people who had to venture outdoors would have some protection. If nothing else, advertising the radiation-protection value of clothing would give advertisers a new slant. Vehicles would need to be to be similarly protected. Air travel would have to be greatly curtailed. As we discussed in chapter 5, the higher up one goes in the atmosphere, the more one is exposed to ionizing radiation. Since this is true even today, the long-term effects of ionizing radiation on airline pilots and cabin attendants is a real issue.

Let's assume, plausibly, that the supernova occurs at a time in the future when people are living on the other planets and moons and in space stations around our solar system. Many of these people would be in grave danger from the supernova because their habitats would lack the earth's thick atmosphere to buffer the effects of the radiation. It is fair to say that the vast majority of the space colonists who aren't living and working deep underground at the time the supernova's electromagnetic radiation arrives would be killed by radiation poisoning, a particularly horrible death.

Many space colonies would become death traps. Unless specially protected from extremes of radiation, space stations and surface buildings on other worlds would be uninhabitable for years. Depending on the self-sufficiency of the underground colonies in Mars, the asteroids, and some of the outer moons, the people living deep out there would face prolonged deaths from starvation or asphyxiation because they could not be resupplied for years. If they ran out of even one essential commodity, such as food, water, air, or a medicine, they would be lost. Therefore, the surviving colonists off-Earth would have to risk a space flight back to Earth. Indeed, our planet is the only place in the solar system with a sufficiently complex biosphere to absorb the dramatic changes brought about by the supernova and still provide a viable habitat for humans.

Long-Term Survival
The long-term challenges facing the human race after the supernova would also be daunting. Considering the radiation hazards on the surface of the earth that would exist for many years, frozen embryos and seed stocks would have to be maintained in shielded environments to assure that healthy animal and plant strains could be reintroduced to the earth when the surface was again safe for wildlife and plants.

The difficulties associated with keeping large herds alive and safe to eat suggests that the consumption of meat would fall dramatically in the years after the supernova. In light of the broken food chains, it is likely that many of the people on the earth who survived the initial radiation poisoning would die in the following decades from starvation and disease.

Long-term survival of the human race would depend, in part, on whether we could coexist with the new plants and animals. Like the poisonous mouse in the example given earlier, other new life might prevent the food chains from being re-established. For example, a new strain of wheat might have the ability to chemically destroy any competing plant life. It might therefore have the capacity of wiping out virtually all crops in the latitudes where it grows.

With geneticists and other scientists struggling to understand the new ecological structure of the natural world, two ethical crises, issues that we are facing in a relatively small way today, would come to the fore. First is the question of defining dangerous or nuisance species and how we should deal with them. Today, for example, some states allow open hunting of certain animals considered to be pests because they "prey" on livestock or domestic animals. Animal-protection groups argue that this is inhumane. At a time when the human race and indeed the rest of nature is struggling for survival, this issue of culling the most threatening species would take on a new meaning. In that case it could be argued that it is "unnatural" to let these breeds survive. But where would we draw the line?

After the supernova, many species of animal would show a range of new physical and behavioral characteristics. Should we hunt down and destroy all those showing even the slightest genetic variation from their ancestors? Do we have the right to do that? If we "allow" some mutant strains to survive, which ones? And where should they live? Should we put them all in zoos or preserves? Should all wild animals be captured for fear that we would otherwise miss one of the "bad" mutants? What happens when mutant strains of insects start destroying crops more effectively? How far should we go in eradicating them or realtering their genetic structures so that the food chains can be re-established? While we do not face the holocaust of a nearby supernova, we are already struggling with a miniature version of this problem. As the human population grows, our need for land increases. We are being forced to decide what lands are to be kept undeveloped so that animals can remain safe in the wild.

The other ethical crisis that would have to be faced by people after the supernova is that of genetic mutations of humans. As with such changes brought about by the supernova's radiation in other animals, most mutant humans would not be viable beings. However, a tiny fraction of these people would survive *to become genetically different from the vast majority of homo sapiens.*

Esoteric as a nearby supernova and its consequences might sound, some estimates suggest they might occur as frequently as once every 25 million years within fifty light-years of Earth. Indeed, the events described in this chapter have probably occurred in the vicinity of the earth at least once, and perhaps several times, since life emerged on dry land. If so, the genetic effects described above may have already happened and changed the course of life on the earth, perhaps nudging it toward the evolution of homo sapiens—us.

WHAT IF A STAR PASSED NEAR THE SOLAR SYSTEM? CERBERON

THE FACT THAT OUR PLANET ORBITS THE SUN AT DISTANCES WHERE water can remain liquid underlies the ability of the earth to support life. The earth ranges between 91.5 million and 94.5 million miles from the sun. Its orbit around the sun would be in the shape of a perfect ellipse if the earth and sun were the only bodies in the solar system. However, the other planets exert different gravitational forces on the earth throughout the year, causing it to swerve from a perfect elliptical orbit by varying amounts at different times. Each time a planet passes closest to us in its orbit, the earth moves slightly toward it.

Fortunately, the effects on the earth's orbit from these other bodies are too weak to move our planet so far away from, or so near to, the sun as to endanger life here. To disturb the earth's orbit that much would require the gravitational force of a much more massive object, such as another star passing closely. In chapter 6 we considered the effects on Earth of extreme quantities of electromagnetic radiation. In this chapter we turn to explore the effects of a major gravitational disturbance. What would happen if a star wandered so close to the solar system that its gravitational force seriously altered the earth's orbit?

In fact, none of the stars near us today will ever come close enough to affect the earth's orbit around the sun. However, other stars could be nudged toward us when the solar system passes through a spiral arm of the Milky Way.

In choosing the intruder's mass and path, we want the earth to be able to support life after the encounter. Two effects from the passing star would be particularly important to consider: its radiation and its gravitation. As with the supernova in chapter 6, too much radiation from the interloper could sterilize the earth's surface. Too much gravitational force from the star would pull the earth out of the region around the sun where life could be sustained. Both of these disasters would occur if an intrinsically low-mass star passed close to the earth or if an intrinsically massive star passed at an even greater distance. What we want for this chapter is a star that will take us through the gates of hell but that won't close them behind us.

PHYSICAL CHARACTERISTICS AND FLIGHT PATH OF THE INTRUDING STAR

We call the intruder Cerberon, after the three-headed dog, Cerberus, who guards the gates of Hades. For reasons that will become apparent shortly, Cerberon is to be a Main Sequence star with half the mass of our sun. Its surface temperature would then be 5,800 degrees Fahrenheit, which is cool as stars go, and it would appear red in our sky. Cerberon would emit one-tenth as much light as the sun.

To give Cerberon a trajectory past the solar system, we begin by noting that the sun orbits around the center of the Milky Way galaxy in a roughly circular orbit at a speed of half a million miles per hour. The solar system is heading toward the present location of the star Deneb in the constellation of Cygnus. (We won't hit Deneb; by the time we get to its present location in 2 million years, that star will have long since moved on.) All the stars in the vicinity of the sun are traveling in the same general direction; none are coming from Cygnus toward us. More plausible than assuming that one will suddenly appear from out of that direction, we assume that Cerberon is charging toward the solar system from the direction of the galaxy's center at the same speed as the solar system is moving around the center.

The earth's average distance from the sun is called an astronomical unit. We set Cerberon on a course that brings it only as close to the sun as the planet Neptune, which orbits at thirty astronomical units. At that distance, Cerberon may seem quite remote, but as we saw in chapter 5 with the small change in the sun's mass, even a modest extra gravitational tug from a star can strongly affect a planet's orbit.

Cerberon's path, combined with its greater distance from the earth than the sun's, would make the intruder appear nine thousand times dimmer than the sun. This is, however, eleven times brighter than the full moon. At the distance of Neptune, Cerberon's gravitational influence on the planets would be important for slightly more than one year. For convenience in describing the interaction between Cerberon and the planets of the solar system, we will assume that the star moves in the same plane as the earth orbits the sun, the plane of the ecliptic.

In keeping with the mythological spirit of three-headed Cerberus, we put two large planets in orbit around Cerberon. This is consistent with current theory that many, if not most, stars with Cerberon's mass have planets. The closer planet will have the same mass as Uranus, while the farther planet will have the same mass as Jupiter. While traveling between stars on its way to the solar system, Cerberon would move in a straight line that wobbles slightly as its two massive planets orbit around it.

The mutual gravitational interaction between the two stars would cause them to change orbits through the Milky Way. Cerberon would follow a hyperbolic path through the solar system, a wide open V-shaped course with the sun at the vertex of the V. Upon leaving the vicinity of the sun, Cerberon would once again move in a straight line, but now in a different direction from its original track. Likewise, after Cerberon's passage, the solar system would no longer be heading toward Cygnus.

Depending on the year of Cerberon's passage through the solar system, some planets would be on the same side of the sun as the intruder and other planets would be on the opposite side of the sun from it. We put Saturn and Mars on the opposite side of the sun. Mercury and Venus, with their orbits taking less than one year, would be on both sides of the sun while Cerberon passed. Pluto, Neptune, Uranus, Jupiter, and Earth will be on the same side of the sun as Cerberon.

At its closest approach we put Pluto directly in Cerberon's path, with Neptune to miss it by two-thirds of an astronomical unit, Uranus to miss it by ten astronomical units, Jupiter to miss it by twenty-five astronomical units, and the earth to miss it by twenty-nine astronomical units.

CERBERON'S IMPACT ON PLUTO, NEPTUNE, URANUS, AND JUPITER

Pluto and its moon Charon would fall directly onto Cerberon, pulled in by the star's gravitational force. Cerberon's heat would vaporize both Pluto and Charon as they fall through the star's outer layers. Compared with Cerberon, Pluto and Charon have such little mass that their entry into it would have absolutely no measurable effect on that star. (Pluto and Charon have a combined mass of less than 1 percent of the earth's mass or less than two millionths of 1 percent the mass of Cerberon.)

Neptune would be torn out of its course around the sun and captured in a permanent orbit around the passing star. Cerberon's heat would then begin vaporizing the gas of Neptune's outer layers. Neptune, along with Uranus, Saturn, and Jupiter, is a gigantic ball of liquid hydrogen and helium surrounding a relatively small, Earth-like core of rocky and metallic elements. None of these planets has a solid surface like the earth or other terrestrial (Earth-like) planets. The hydrogen and helium comprising Neptune's surface would be heated sufficiently by radiation from Cerberon to drive these gases off the planet and into space. As a result, Neptune would lose mass and shrink until nothing but its terrestrial core was left.

Heated by Cerberon, the core would start giving off some of the carbon dioxide and water locked in its rocks. These would form an atmosphere around the remnant core. As the core cooled, the water vapor in its air would condense onto its surface, forming oceans. The carbon dioxide would be absorbed into the oceans (sound familiar yet?), and lightning and ultraviolet radiation would create basic biochemical building blocks of life, which would rain down into the oceans. Water from the land would also stream into the oceans, carrying sediment that would provide raw materials for more life to form. Protected from ultraviolet radiation from Cerberon by several yards of water, life would evolve in the oceans and eventually emerge onto the dry land of a planet that had once been a ball of liquid.

Uranus would feel twice as much gravitational attraction from Cerberon as it does from the sun. As a result, it would be pulled out of orbit around the sun. But Uranus would be too far away from Cerberon to be captured by it.

Instead, it would be cast adrift, doomed to travel forever in the frigid, black silence of interstellar space.

Jupiter has the misfortune of being in the same place at the same time as Cerberon's farthest planet. The two bodies would collide and, because they have the same mass, destroy each other, shedding the bulk of their mass. Being closer to the sun than to Cerberon, most of the hydrogen and helium released from the two planets would form a ring or cloud around the sun. Over time this gas would be pushed out of orbit and out of the solar system by impact from radiation and particles coming from the sun. The terrestrial cores of Jupiter and the planet that strikes it would also be pulverized in the collision. Depending on the details of the collision, some of this rubble would remain in orbits around the sun and Cerberon. Some of this debris would be knocked toward the inner solar system. It would only be through extreme luck or diligence on the part of the people then living if the earth were to avoid being struck. Those people would have to deflect the incoming debris with explosives.

CERBERON'S IMPACT ON THE EARTH AND MOON

Because Cerberon is to pass at twenty-nine astronomical units from Earth, the star's electromagnetic radiation would have a negligible effect on our planet. While Cerberon would be bright in the sky, its ultraviolet radiation would not profoundly damage the ozone layer, nor would its infrared radiation heat the earth noticeably. On the other hand, Cerberon's gravitational force on the earth, ten times greater than Jupiter's attraction on our planet, would have to be reckoned with. Cerberon's force would be about seventeen hundred times less than the sun's pull on us, yet it would be great enough to significantly alter the earth's orbit and the fate of life here.

Details of how Earth's orbit would change during Cerberon's tenure in the solar system depend sensitively on the relative speeds between the earth, sun, and Cerberon. If the earth were heading away from Cerberon as the star entered the solar system, it would be slowed down by the star. As a result, the earth would move into an orbit closer to the sun. If the earth were directly between the sun and Cerberon at Cerberon's closest approach, the intruder's gravitational force would pull it outward in its orbit. The greatest change in the earth's orbit would occur if Earth were coming around the sun toward Cerberon. In that case the star would strongly accelerate the earth away from the sun. This situation is called a resonance between the planet and Cerberon.

The principle of resonance is similar to that of pushing a child on a swing—if you push at just the right time, you can make the swing go higher and higher. If you push at the wrong time, you slow the swing down. In the worst case, the earth could be pulled out of the zone around our sun in which life can be supported.

Of all the possibilities for gravitational interactions between Earth and Cerberon, we choose to have the intruder pull the earth into a more elliptical orbit while leaving it in an inhabitable range of distances from the sun. After this encounter, the earth's farthest point from the sun (aphelion) would be farther out than it is now, while its nearest point to the sun (perihelion) would be closer. This restriction on the change in the earth's orbit may seem to take some of the fun out of such an encounter, but at least it leaves us alive. Besides, the other things we would have to deal with are more interesting than being cast adrift between the stars like Uranus.

The moon's orbit around the earth would be changed by Cerberon's gravitational attraction. As Cerberon passes through the solar system, the direction in which it pulls the moon would vary. This, combined with the moon's changing position relative to the earth during the encounter, would cause the moon to end up in a more elliptical orbit around the earth.

EARTH GEOLOGY AS CERBERON PASSES THROUGH THE SOLAR SYSTEM

Pulling the earth out of the orbit it has had for over 4.5 billion years would disrupt the fragile equilibrium established between the planet's molten interior and its thin, brittle crust. The interior would begin to slosh around as it unexpectedly changed paths under the gravitational influence of Cerberon. This motion would breach the crust, creating earthquakes and activating volcanoes in numbers not seen here in billions of years. These activities would create tsunamis (tidal waves) all around the globe.

While Cerberon was passing, the displacement of the moon from its present orbit would alter the moon's tidal force on the earth. This, in turn, would also generate massive tidal waves as the oceans tried to readjust to the changes in the moon's position. The tremendous churning of the oceans would also create powerful storms as heat stored in the water is brought to the surface and released. Therefore, the passage of Cerberon would also be notable for extremely unsettled weather.

The volcanoes and storms would actually have partially competing and partially compensating effects on the climate during this period. As we saw in chapter 1, volcanoes put tremendous amounts of gas and dust into the atmosphere. The dust would prevent some of the sunlight that normally reaches the surface from getting there. This would add to the cooling created by increased storm activity, since the tops of clouds reflect more sunlight into space than usual. In that way the two phenomena reinforce each other.

On the other hand, the torrential rains associated with the storms would carry down with them large quantities of the dust that had been injected into the atmosphere by the volcanoes. Without the storms to wash the sky clear, the dust would probably keep the surface dangerously cold for decades. As it is, the year or two of storms associated with Cerberon's passage should remove much of the dust, leaving the surface temperature relatively tolerable within a few years. In that way the storms partially counteract the effects of the volcanoes.

CHANGES TO THE EARTH'S INTERIOR

The quake and volcano activity would begin to subside after Cerberon departs and the earth becomes established in its more oval-shaped orbit around the sun. However, our planet would remain more seismically active than normal for hundreds, possibly thousands, of years as its turbulent interior settles down.

One possible outcome of the encounter with Cerberon on the earth's interior is a dramatic change in the direction of the earth's magnetic field. While geologists do not completely understand the origin of the magnetic field, evidence of its changing orientation is found in rock on the earth's surface and at the bottoms of the oceans. Three things are evident. First, the earth's magnetic field rarely pokes through the earth's surface at the north and south rotation poles. Second, the places where the magnetic poles emerge from the earth's interior change; that is, the earth's magnetic field wanders. For example, over the past 150 years it has moved about eight degrees closer to the rotation poles. And third, the field sometimes flips completely so that the north magnetic pole emerges from the earth near the south rotation pole and vice versa.

Plausibly, the disorder created inside the earth by Cerberon would cause a significant dislocation of the earth's magnetic field, perhaps even causing it to flip from north to south. As a result, the Van Allen belts would become disrupted. (They disappear completely when the flips occur.) As a result of the weakening of the field, more particles from the solar wind would enter the earth's atmosphere, creating more frequent auroras as well as damaging the ozone layer. More ultraviolet radiation from the sun would reach the earth's surface for years after the fields are disrupted until the field and Van Allen belts restabilize.

A CHANGE OF SEASONS

The most important effect of the increased ellipticity of the earth's orbit would be on global weather patterns. In chapter 4 we saw that while the earth's orbit today is measurably elliptical, the weather is not greatly affected by the change in distance from the sun. This is because the two hemispheres of the earth have different areas covered by land and ocean. The greater ocean area in the Southern Hemisphere has a slight cooling effect that compensates for the slight increase in heating during summers down there. Therefore, even though the earth is closer to the sun in January, the southern summers are not noticeably hotter than those enjoyed in the Northern Hemisphere.

The seasons in the two hemispheres would, however, become significantly different after the intruder's passage. In particular, the seasons would become more severe in one hemisphere, while being milder in the other. Which hemisphere receives the benefit and which the harm would depend on the tilt of the earth's axis. In chapter 4 we also saw that the seasons are primarily determined by how much the earth's axis is tilted in relation to the ecliptic. Throughout the year, this tilt causes the height that the sun rises in the sky and the number of daylight hours to vary. To understand the seasons after Cerberon passes, we will add to the effect of the tilted axis the greater change in heating of the earth as its distance from the sun varies throughout the year.

The earth is to be farthest from the sun during summers in the Northern Hemisphere. This means that when the North Pole is inclined toward the sun, the earth is to be at aphelion. (The same is true today, but as we have just seen, the effects of distance are insignificant.) After Cerberon passes, northern summers would therefore be cooler than they are today. During winters in the Northern Hemisphere, when Earth would be closest to the sun, temperatures would be warmer than they are now.

Conversely, the earth's Southern Hemisphere would be closest to its sun during its summers, making that season even hotter, while its winters would be even colder than they are today. Therefore, the greater difference between the earth's perihelion and aphelion distances from the sun would lead to a significant difference in seasonal temperatures in the northern and southern hemispheres. This, in turn, would change the habitability of the two hemispheres as well, which, of course, would lead to changes in which types of plant and animal life exist in the different hemispheres. We will have more to say about that shortly.

The change in distance from the sun would also affect the ice cover in the earth's polar regions. As the Southern Hemisphere sweltered during its hotter summers, some of the ice on Antarctica would begin melting and running into the ocean. Likewise, more ice in the ice shelves surrounding that continent would break off, creating icebergs that would float northward. While winter snowfall would replenish some of the ice, the average ice coverage on Antarctica would be lower than it is today. Indeed, after Cerberon's passage there may well be large bare regions on Antarctica, areas that are covered with thick layers of ice and snow today. Conversely, more ice would build up in the Arctic Ocean around the North Pole.

The complexity of the interactions between the oceans and atmospheres makes the final ice coverage of the new earth impossible to predict. It would depend on the details of the earth's new orbit. Hidden in this change is the possibility that increased ice coverage in the Arctic and snow coverage on the continents of the Northern Hemisphere would lead to runaway glaciation, as discussed in chapter 5. Happily, this problem would be averted, at least for thousands if not millions of years, by the vast quantities of carbon dioxide gas ejected into the atmosphere during the volcanic eruptions generated by Cerberon's passage.

Recall that carbon dioxide is a greenhouse gas that stores heat. Volcanoes and fissures in the earth's early surface were responsible for the original carbon dioxide put into the atmosphere. There are still enormous reserves of such gas inside the earth. With the new cycle of volcanoes active during Cerberon's presence, more carbon dioxide would be released into the air. This gas would keep the air warmer than it would otherwise be during the times of northern winters. Therefore, less ice and snow would be deposited on the earth's surface after the encounter, and runaway glaciation would probably not occur.

But the threat of a runaway greenhouse effect created by the greater amount of heat stored in the atmosphere by the carbon dioxide would grow. As indicated in chapter 5, the earth today is balanced between the two runaway effects. Without a doubt, that balance would be upset by Cerberon. For the good of life on Earth, we hope that the new equilibrium climate established after Cerberon is long gone falls somewhere between runaway glaciation and the runaway greenhouse effect.

DEALING WITH COMETS

Like the sun, Cerberon would escort an enormous swarm of comets around the galaxy. Orbiting billions of miles from the star, some of these frozen bodies would precede it into the solar system, and some would be here years after it departed. Some of Cerberon's comets would be captured by the sun, while some of the sun's comets would be pulled out of orbit to become part of Cerberon's host. The earth and its inhabitants would face grave danger from Cerberon's comets plunging into the inner solar system, pulled by the sun's gravitational force.

The comets drawn sunward during Cerberon's intrusion would develop tails as they became heated by the sun. They would appear as diaphanous streaks across the night sky for decades before and after Cerberon's actual appearance here. The problem is that the earth is sufficiently close to the sun and has sufficiently strong gravity that it would almost certainly pull onto its surface one or more of Cerberon's comets. This concept of impact from space is presently the most plausible explanation for the extinction of the dinosaurs 65 million years ago. Based on the paleontological and geological evidence of past impacts, it seems likely that a comet's impact could damage the earth's atmosphere, surface, and oceans as well as annihilating half the species of animal and plant life on Earth.

While it would be technically impossible to change Cerberon's course toward the solar system so as to avoid all the changes it would wreak, we already have the capability of deflecting comets so they do not hit the earth. As with deflecting the parts of the destroyed Jupiter drawn toward the earth mentioned earlier, small nuclear charges could be implanted on an incoming comet and then detonated. Correctly placed, the explosion would apply a force perpendicular to the comet's path, pushing it into a new orbit and away from the earth. This must be done with some care because comets are small enough so that detonating a large nuclear bomb on one could pulverize it, leaving some of the pieces to plunge onto the earth. Nevertheless, the potential effects of large comet impacts resulting from Cerberon's presence could be averted by technology.

CHANGES ON EARTH DUE TO
THE MOON'S NEW ORBIT

A new lunar orbit means a change in the length of time it takes the moon to orbit the earth. This would change the interval between high tides and the time of a complete cycle of lunar phases. Because the moon's orbit would be more elliptical, the height of the ocean tides would vary a great deal throughout the new lunar month, since tides are very sensitive to the distance between the earth and moon.

LIFE ON EARTH DURING AND AFTER
CERBERON'S PASSAGE

Like the earth's bombardment by supernova radiation, the passage of Cerberon would cause major dislocations in the earth's food chains, many of which would never be re-established. Tidal waves during the period of encounter would destroy the habitats of many shorebirds, turtles, and sea mammals such as seals and sea lions. Coastal nesting sites would be unusable, and many aquatic creatures and coastal plant life would quickly become extinct. More flexible species would need to seek safer nesting and feeding areas.

The volcanoes would cover large areas of land with lava, killing all the life there and rendering the land uninhabitable for several years. The cooling of the planet due to the volcanic emissions would immediately wipe out species unable to migrate or adapt to the change in climate. Other species would need to relocate to warmer climes and compete for territory with animals already established there. This, too, would lead to the loss of many species. The carbon dioxide emitted into the air from volcanoes would then heat the earth, further straining the abilities of different types of plants and animals to adapt and survive.

To make the weather and tidal changes even worse, the moon would be changing its orbit, as mentioned previously.

The greater variation in its distance would lead to a greater range of tides than it now creates. High tides would bring the oceans much farther inland than they are today. This would increase the available area of the intertidal region for those life forms that function well there.

The end of the star's passage through the solar system would herald the beginning of a new era for the earth and its inhabitants. The worldwide traumas created by the gravitational pull of the star would subside, and the earth and moon would become established in their new orbits. All life would begin adapting to new weather patterns around the world.

Once the seas and weather have settled into a new equilibrium, tidal waves and storms would subside. More coastal land would become inhabitable again. Animals, compressed into smaller regions of the earth during the star's passage, would spread back over the earth's surface, but not necessarily to their ancestral homelands. The major point here is that many of the niches for life throughout the earth would no longer be suitable for those plants and animals that previously lived in them. To a large extent, plants and animals would play musical habitats, finding regions that were previously uninhabitable for them now to be suitable.

For years, perhaps decades, the animals of the world would compete tooth and fang for new territory. Without a doubt, their most dangerous foes would be human beings. The problems of all animals would be greatly amplified if years of crop failure also occur during Cerberon's passage. Then, as will discuss in more detail, we humans would be in need of more meat for food, perhaps thereby endangering with extinction species that otherwise would have survived Cerberon's presence in the solar system.

The first new equilibrium established in the animal kingdom after Cerberon's departure would not necessarily be the last. Even as the remaining creatures on Earth struggle to find new niches for themselves and establish new food chains, subtleties of weather, disease, new and varied competitors, and other factors would make necessary additional adjustments in the relationships between the animals of the world. Further reshuffling of territories for plants and animals would occur; different food chains would develop. This process of readjustment to available food and new climates would continue for centuries.

HUMAN DISLOCATION DURING CERBERON'S PASSAGE

It is reasonable to expect that all of the coastal cities of the earth would be heavily damaged by wind, waves, earthquakes, and volcanoes during Cerberon's passage. During that time they would not be inhabitable by humans, leading to the largest series of worldwide migrations imaginable. Many inland cites would also be damaged by earthquakes and volcanic eruptions. Everyone in the world who survives the immediate destruction would face unparalleled hardship in procuring even the basic necessities of survival. Their problem stems from the global community that the earth has become.

Virtually every city, town, and village on Earth is dependent for its survival on products from many other places. These goods include fuel, food, consumer products, building materials, and medical supplies, among innumerable others. Many of these, especially food, petroleum products, and medical supplies, are usually not stockpiled locally in large quantity. For example, if all food shipments to a small city or town abruptly stopped one day, the residents there would begin experiencing food shortages within a week. While the stoppage of goods would not occur overnight during Cerberon's passage, virtually all transportation routes would eventually be cut off by the tides, storms, and earthquakes. Even stockpiled goods would run out long before reliable land, sea, or air travel could be resumed.

Consider the case of ocean shipping. High tides, tsunamis, and powerful storms would prevent the travel of virtually all ships. In fact, we could expect most ships to be sunk since these natural phenomena would be powerful enough to penetrate even the most secluded harbors. Some of the crucial products transported worldwide in large quantity by ships today are oil, natural gas, and food grains. Many countries do not have these products in their territories and are completely dependent on ships to supply them. The people of island nations such as Japan, the Philippines, Taiwan, Sri Lanka, and Madagascar would be most at risk during this period of Cerberon's presence.

The alternative routes for goods normally shipped by sea would not be much more viable. Land routes would often be disrupted by earthquakes and lava flows. The air would often be filled with precipitation and dust, making transport flights all but impossible. For a few years long-distance transportation would literally and figuratively grind to a halt.

Most people in northern cities (of the Northern Hemisphere) are dependent on oil and natural gas to heat their homes. Without alternative energy sources of heat during the winter, these people would have to migrate south en masse, perhaps returning home in the spring. While heating problems would most affect people at higher latitudes, difficulties getting food would affect everyone. Food shortages would cause further migrations.

Much of the worldwide electrical grid would also be interrupted by Cerberon's passage. Those power plants fed by oil, coal, or natural gas would run out of fuel during the year of Cerberon's presence in the solar system. Power stations on the lower reaches of rivers would be damaged by devastatingly high tides and waves channeled up toward them from the ocean. The loss of electrical power would force further displacement of the population.

One could go to considerable lengths in this vein, but it would take us too far afield. Suffice it to say that the earth would rejuvenate as after the other great periods of dislocation for life (for example, when space debris struck the planet). Perhaps the most important lesson to be learned from considering such encounters as the passage of Cerberon is that if the human population becomes too large, the demands of people during global crises could lead to the annihilation of other species. Indeed, even today our needs lead to the destruction of animal and plant habitats. Ideally, the human population, postencounter, would not grow as large as it is today, thereby keeping the world more fit for a balanced existence between humans and the rest of nature.

WHAT IF A BLACK HOLE PASSED THROUGH THE EARTH?
DIABLO

IN THE LAST TWO CHAPTERS WE HAVE SATURATED THE EARTH WITH radiation and dragged the planet into a new orbit, respectively. Antar in chapter 6 and Cerberon in chapter 7 both acted on the earth from a distance. In this chapter we are going to throw something right at it and see how the earth responds. We already know what would happen to the earth if a terrestrial planet the size of Mars were to strike it. We saw in chapter 1 that such an impact would splash the moon into orbit. Here we want to consider the effects on Earth of something more exotic, more ominous, more unstoppable: a black hole.

PROPERTIES OF BLACK HOLES

The concept of a black hole arose in its present form in Albert Einstein's general theory of relativity, first enunciated in 1915. This theory predicts, among many other things, that whenever any amount of matter is compressed beyond a critical density, it fundamentally alters the space around it. Empty space near this dense mass is actually bent or curved so that all matter trying to escape from this region is led back into it.

To escape from the earth's gravitational grasp and enter free space, a projectile fired from the earth's surface has to be moving at nearly seven miles per *second* as it begins rising. To escape from our moon's gravitational field requires an initial velocity of only one and a half miles per second, or five thousand miles per hour. From Jupiter a projectile would need an initial speed of 37 miles per second, or 130,000 miles an hour. A projectile fired directly upward from anywhere inside a black hole would have to be traveling faster than the speed of light—186,000 miles per second—in order to escape from it. However, according to our current understanding of the laws of nature, no massive object can move at or beyond the speed of light. (As matter speeds up, its mass automatically increases. This proven effect from Einstein's theory of special relativity, in turn, makes it harder to speed the matter up further. It would take more energy than the universe has to move even a dust mote up to the speed of light, at which point its mass would be infinite.)

Similarly, even photons fired straight out would lose all their energy before they got to the black hole's outer boundary, called the Schwarzschild radius. Whereas projectiles are given an initial boost and left on their own, you might wonder whether a rocket firing continually could lift itself out of a black hole. In fact, it could not. This is not an engineering issue; there is just not enough energy (rocket fuel) in the entire universe to lift even the smallest rocket out of a black hole's gravitational field.

The matter in the black hole does not fill the space out to the Schwarzschild radius. Indeed, because the matter in the black hole is so dense, there is no known force in nature that can prevent it from continuing to collapse in on itself forever. Therefore, once compressed into being a black hole, the matter continues to collapse for eternity! This matter becomes concentrated either at the center, if the hole is not rotating, or in a ring around the center, if it is rotating.

Black holes are believed to form under at least two distinct circumstances. The first situation is as the end-state of the collapse of massive stars, as we discussed in chapter 6. The other circumstance is from the compression of tiny amounts of matter during the explosion that created the universe in the first place, the Big Bang. In this chapter we will throw both kinds of black holes at the earth, like lightning bolts from Zeus. Since we have already begun describing the process of creating a black hole from a star, we will hurl one of these first.

COLLISION WITH A
STELLAR-REMNANT BLACK HOLE

Origin of the Black Hole

Stellar-remnant black holes form from the cores of massive stars. The collapsing neutron star in chapter 6 would become a black hole. Since the neutron star has no outward force capable of withstanding the inward pressure from its own gravity, it would continue to condense. All four solar masses of neutrons from the collapsing neutron star would become so concentrated that they would plunge inside their Schwarzschild radius, thereby becoming a black hole. The collapse would not stop there; the black hole's own gravitational force would make the neutrons exert so much pressure on one another that they would finally coalesce into a form of matter that exists nowhere else.

The new state of the black hole's mass would retain only three recognizable properties: its mass, its total electric charge, and its angular momentum. The electric charge is the difference in the number of protons and electrons that went into forming the hole. Most black holes are believed to have come from stars containing equal numbers of these particles. We therefore assume that both of the black holes that strike the earth in this chapter are electrically neutral.

Angular momentum measures how fast the black hole is rotating. If the star that forms a black hole rotates—as do most stars, the sun included—the black hole would rotate. Since the star from which our black hole came was rotating, so is our black hole. Rotating black holes are called Kerr black holes, after R. Kerr, who first described them mathematically in 1963.

As a result of the continual collapse of the black hole's mass into a thin ring, there is no solid surface at the Schwarzschild radius of the black hole. In fact, the Schwarzschild radius is not marked by any special characteristic to let you know you've passed it. It is just empty space. Only when you find it impossible to leave would you know you were inside the Schwarzschild radius.

Once formed, the Schwarzschild radius is at a fixed distance from the center of the hole even though the matter in it continues to contract. The exact distance from the center of the black hole to its Schwarzschild radius is determined uniquely by the mass contained within the black hole. *Our four-solar-mass black hole has a Schwarzschild radius of just seven and a third miles.*

Contrary to the fears of many people, once a black hole is formed, it does not suddenly blossom into an all-consuming monster gobbling up everything around it. A spaceship just a few thousand miles from the black hole would feel only the gravitational pull of the four solar masses of matter in the hole. Granted, such gravitational attraction is enormous, but the black hole would not act as a vacuum cleaner pulling the ship into it with more than just its gravitational force. With sufficient power, that ship could orbit the black hole and then blast away from it without any adverse effects to its occupants.

Discovery of the Incoming Stellar-Remnant Black Hole

Astronomers today can discover only black holes that are part of binary star systems. Binary systems are pairs of stars orbiting around each other. Such couplings are quite common, occurring as often as isolated stars. A very massive star in a binary system can supernova without destroying its companion. In such situations, the resulting black hole will pull some of its companion's upper gas layer off and toward it. But since the black hole is only a few miles across, much of the infalling gas has to wait its turn before passing through the Schwarzschild radius. This gas begins orbiting the black hole, creating an accretion disk.

We observe the black hole because of its accretion disk. Gas on the inner edge of the disk swirls in through the black hole's Schwarzschild radius like water going down the drain of a bathtub. In the meantime, more matter from the companion star heads toward the black hole. At high speed, this fresh gas strikes the accretion disk somewhere outside the Schwarzschild radius. The resulting impact between the gas of the disk and the incoming gas creates X rays that are radiated into space. The signature of a black hole to astronomers today, then, is a binary star system in which one member, containing more than about three solar masses, is invisible and gives off X rays. Such systems have been observed in several constellations including Cygnus (the Swan) and Hercules.

The black hole heading toward the solar system is to come alone. And yet astronomers might well learn of its approach from electromagnetic radiation it emits in a process related to the accretion-disk X rays of a black hole in a binary star system. Our black hole could not avoid passing through some of the ubiquitous interstellar gas and dust on its journey toward the solar system. Since the black hole is so small compared to the volume of gas it would attract, much of the gas attracted to it would have to wait its turn to fall inside. The gas would swirl around the black hole, becoming compressed as it descends toward the Schwarzschild radius. The compression of the gas would generate electromagnetic radiation, either X rays or a lower-energy variety such as ultraviolet radiation. These photons could be detected long before the black hole entered the solar system.

The process of determining whether this peculiar source of radiation is an incoming black hole would follow the normal course of events associated with discovering a new source of radiation in space. The types of radiation emitted by different astronomical objects and events are all different, which is why radiation is referred to as a signature of its source. Upon discovering the black hole's emissions, astronomers would first try to classify it among known radiators. After eliminating all known sources, astrophysicists would think up new ones, determine what their signatures should look like, and compare these results with the new radiation. Inevitably, if only because every other source proved incompatible with the observed radiation, someone is bound to suggest a black hole approaching the solar system.

Knowing that a black hole is coming and doing something about it are two different matters. Whereas it is technologically possible to deflect comets headed toward the earth (see chapter 7), there is no conceivable way to deflect a four-solar-mass black hole. After all, despite its tiny size, this object has more mass than the sun! As the massive black hole sweeps through the solar system, planets near it would be pulled out of their normal orbits by its gravitational force. Some of them would be cast out of the solar system. Others would be sent into much more elliptical orbits around the sun or into orbits around the hole. We assume that none of the other planets has the misfortune of being pulled directly to the black hole; we leave that tragedy to the earth alone.

Encounter with the Stellar-Remnant Black Hole

As in chapter 7, the earth's first response to the black hole would be to drift out of its present orbit. This would be accompanied by earthquakes and volcanoes, which would create the same havoc as did Cerberon's presence. These, however, would be the least of the earth's problems. As the black hole draws closer, the earth would begin to feel a tidal force from it similar to that exerted by the moon on the earth. Unlike the moon's tidal force, however, the one from the black hole would continue to grow as it approaches us. It would become strong enough to cause the land closest to the approaching black hole to rise up, as the oceans do toward the moon today.

Recall from chapter 2 that for every planet there is a minimum distance at which the moon must be located in order to avoid being pulled apart by the planet's powerful tidal force. Closer than this so-called Roche limit, the moon would be torn to pieces. The same thing would happen to the earth as the black hole approached. The earth's surface would already be flying apart by the time the black hole is within 400,000 miles of the earth (twice the distance from the earth to the moon).

After stripping away the earth's crust, the black hole would start to pull out its red-hot mantle, exposing the earth's white-hot core. As the black hole neared it, even the earth's core would disintegrate until all that was left of the planet would be so much hot, dispersed space rubble. The black hole would pass through the debris that had been the earth, pulling into itself those parts of the destroyed planet that happen to be inside its Schwarzschild radius. The black hole's gravity would drag much of the remaining matter from the earth behind it.

Considering how much mass it would have compared with the earth, the black hole's orbit and mass would be virtually unchanged as a result of the encounter. The hole would continue on its way through the solar system, destroying as it goes. Eventually it would move beyond the solar system, never to return.

What little mass of the earth that remained in orbit around the sun would be spread too far and wide to ever recondense into a single body. Rather, the debris left here would orbit the sun as a new asteroid belt. Unless the humans on Earth during this period already have large interstellar spaceships and have left the earth before the black hole gets too close, all terrestrial life would be exterminated.

COLLISION WITH A PRIMORDIAL BLACK HOLE

Undaunted by this depressing state of affairs, we now turn our attention to a collision between the earth and a black hole formed at the beginning of the universe, called a primordial black hole. Theory has it that some black holes were created during the first few seconds after the Big Bang, the explosion that started all the matter and energy of the universe expanding outward from an initially microscopic volume. This expansion is observed to be continuing today. Each primordial (or mini-) black hole represents a tiny clump of matter compressed inside its Schwarzschild radii by the force of the Big Bang explosion around it. Happily, primordial black holes are believed to contain much less mass than the stellar-remnant holes just described. If they exist, primordial black holes are believed to have formed with masses that range from ounces to more mass than is contained in the earth. Therefore, a primordial black hole would do much less damage to the earth.

There is a lower boundary to the mass that a primordial

black hole passing through the earth could have today, which is greater than the minimum mass of a few ounces for a primordial black hole. The difference is due to the fact that the lowest-mass primordial black holes are believed to have already dissipated. To see how this could happen and to help us select the mass of the primordial black hole passing through the earth, we consider a mechanism for removing black holes from the universe, called the Hawking process.

The Hawking Process

One intriguing proposal suggests that all of the tiniest primordial black holes should have already vanished from the universe. This idea, put forward by and named for renowned British astrophysicist Stephen Hawking, says that *all* black holes actually evaporate and that the smallest primordial black holes do so most rapidly.

It is not intuitively obvious that black holes should be able to lose mass (evaporate). And given that they can, it is even less obvious that the holes with the lowest mass (and therefore the least gravitational force) should do so faster than the more massive ones. If no matter or light can pass out through a black hole's Schwarzschild radius, how can the black hole lose mass? The Hawking process invokes a subtle but experimentally established phenomenon, called virtual-particle or virtual-pair production, that occurs just outside the Schwarzschild radius.

Virtual-particle production describes the formation of pairs of particles that occurs spontaneously everywhere in the universe. Under normal circumstances, such as in your body as you are reading this book, the particles have no effect on anything because they annihilate each other so quickly. Virtual-particle pairs appear and then disappear in less than a billionth of a billionth of a second. Because they exist for such short intervals, the appearance of these particles does not violate any laws of nature, such as conservation of mass-energy (which says that you cannot create matter or energy out of nothing).

Virtual particles are always created in pairs matching a particle and its antiparticle, such as an electron and a positron or a proton and an antiproton. (Particles and antiparticles are exactly the same particles except that they have opposite charges. Both types of particles have *attractive* gravitational forces.) Virtual-pair production is not just a theoretical construct or a prediction of the equations of quantum mechanics (which describe the dynamics of particles). These particles have been observed in particle accelerators, when high-speed particles in the accelerator beam inadvertently strike a virtual-particle pair before they have time to annihilate each other. The impact separates the virtual particles, making them permanent members of the universe. The cost of doing this is that energy is removed from the particle that struck them.

Let us return to black holes. Immediately outside their Schwarzschild radii, virtual particles form, just as they do everywhere else in the universe. The pairs of virtual particles created just outside a Schwarzschild radius feel the tremendous gravitational force of the hole pulling on them. Sometimes virtual pairs form with one particle closer to the black hole than the other. In that case, the hole exerts a different amount of gravitational force on the two particles, exactly as the moon exerts different amounts of force on different places on the earth. If the *difference* in forces felt by the two virtual particles near a black hole is great enough, the particles will be pulled apart before they can annihilate each other. As a result they are made into real particles. Just like the moon's force on the earth, the force ripping the two virtual particles apart and making them real is a tidal force.

Of the two freshly made real particles, the one closer to

the black hole always falls into it, while the other particle can avoid falling in if it is made real with enough energy to fly away. If the second particle moves outward faster than the escape velocity at the black hole's Schwarzschild radius, it ascends like a projectile into free space.

It is in this situation, in which one newly formed particle moves outward with enough energy to escape into free space, that black holes evaporate. This particle contains mass that it acquired at the black hole's expense. The energy needed to make the two virtual particles real comes from the black hole's gravitational force, which naturally extends outside the black hole. It happens this way: The gravitational energy that went into making the virtual particles real was created by the mass inside the black hole and transmitted outside as part of the black hole's normal gravitational attraction. This energy was then permanently removed from the gravitational field of the black hole (and therefore from the black hole itself) in the form of the real particle traveling away.

As a result of the Hawking process, the black hole permanently transforms an amount of energy into real matter. The energy loss, E, by the black hole is given by Einstein's equation $E = mc^2$, where m is the total mass of the two particles created and c is the speed of light. The fact that some of the newly formed mass permanently leaves the black hole's vicinity means that the black hole permanently loses that much mass. This causes the black hole to evaporate. Therefore, the hole loses mass not by having any matter pass through its Schwarzschild radius but by having the gravitational energy that does extend outside it transfer its own mass out in the form of energy.

But doesn't the one particle falling inward increase the black hole's total mass? Why doesn't the black hole grow rather than evaporate? The one infalling particle does indeed increase the black hole's mass. But because the black hole gave up the mass of two particles when it made the virtual particles permanent, it loses more than it gains from the one incoming particle.

In principle, we can detect the newly created particles flying away from the vicinity of the black hole. These electrons, positrons, protons, antiprotons, or other fundamental particles should seem to appear from nothing, since we can't see the hole itself. Such spontaneous fountains of particles have yet to be discovered.

Theoretically, all black holes are continually losing mass through the Hawking process. The smaller and less massive a black hole is, the faster it creates particles. This occurs because the closer you are to the center of a black hole, the greater the tidal forces are. Even though a less massive black hole has less total gravitational force, its smaller Schwarzschild radius allows the virtual particles formed near it to actually feel a greater tidal force than those formed around more massive, but physically larger, black holes. Therefore, more of the virtual particles around a less massive black hole are made real every second than around a more massive black hole; the lower-mass black hole evaporates more rapidly.

The smallest primordial black holes should have already lost all their mass through the Hawking process. That is, they should have already vanished. In the final stages of evaporating, tiny black holes create real particles so quickly that they may even appear to be exploding. As a result of Hawking evaporation, all black holes of less than about 1 trillion pounds (if weighed on the earth) should have vanished by now. Therefore, any black hole passing through the earth today would have more than this mass.

The upper limit to the mass of the primordial black hole of interest here is one that would tear the earth apart, as did the stellar-remnant black hole. A reasonable estimate of this upper limit is a black hole with the mass equivalent to that of the earth. Accordingly, we allow the primordial black hole that passes through the earth to have the same mass as our moon. The Schwarzschild radius for this black hole is only 1 ten-thousandth of an inch from its center! Furthermore, we assume that the black hole has no charge and that it is rotating (a Kerr black hole).

Detecting the Primordial Black Hole

A black hole with the mass of the moon would attract little interstellar gas. We will call this intruder Diablo. Since it would radiate very little, if at all, astronomers would not know of its presence until after it entered the solar system. In the unlikely event that Diablo passed close to an asteroid or a moon of one of the outer planets, astronomers could learn of its presence before it arrived at the earth by observing its gravitational effects on these objects. But considering the size of the solar system, the likelihood of the black hole passing close enough to measurably deflect any of the outer planets or an asteroid before reaching the earth is slim. In all likelihood, we would probably first know of Diablo only a few hours before it struck the earth. (Since astronomers do not yet know whether primordial black holes exist and, if so, how many there are and where they are located, we cannot yet calculate the likelihood of this event occurring.)

Encounter with a Primordial Black Hole

The amount of damage done to the earth by Diablo would depend on the speed at which it passes through the planet. Like its mass, its slowest speed into the solar system can be readily determined. As the hole falls toward the inner solar system from interstellar space, it would pick up speed, just as a comet does after being deflected inward from the Oort comet cloud. Both objects are converting some of the gravitational potential energy they feel from the sun into energy of motion, called kinetic energy. (The same thing happens when an object is dropped on the earth.)

After reaching their closest approach to the sun, both the comet and the black hole return to the position at which they first began falling sunward from the solar system, unless they lose energy due to some outside influence along the way. In particular, an unimpeded black hole would gain so much speed (kinetic energy) falling in through the solar system that it would be able to completely escape from it. Another way to say this is that a black hole that doesn't lose energy always has a higher speed than the escape velocity from the solar system. On its way in toward Earth, Diablo meets this criterion. We set the black hole's approach speed to Earth at 100,000 miles per hour. This is only slightly greater than the 94,000 mile-per-hour velocity needed to escape from the solar system when starting outward from the vicinity of the earth.

We assume that Diablo approaches from the side of the earth opposite the moon. By the time the black hole is within the moon's distance from Earth (230,000 miles), the oceans here would begin to feel a tidal force from it and would rise in response. However, moving at 100,000 miles an hour toward the earth, Diablo would be approaching so rapidly that the oceans and (eventually) the solid parts of the earth would not have time to rise very far before it struck the planet. Indeed, during the last two hours or so before impact, the hole's gravitational force would enhance the high tide, but the water would barely have time to start serious flooding, and the earthquakes and volcanic activity would have only just begun as it struck.

As the black hole enters the earth, the gravitational force it would exert around the area of impact would be enough to disrupt the land. In other words, the surface of the earth close to ground zero would be dislocated by momentarily being pulled upward toward the black hole and then dragged violently back downward as the black hole pierced the surface. The earth around the entry and exit points would be so stressed by Diablo's gravitational force as to create weak spots on the surface through which lava would quickly emerge. This activity would be accompanied by violent earthquakes, which you could visualize as larger versions of the ripples created by throwing stones into a pond.

It is only because the black hole would be moving so rapidly that the entire surface of the earth would not be completely disrupted by its gravitational force. Rather than slowly pulling the earth apart, our Kerr black hole has time only to hammer it. Indeed, Diablo would cross through the entire earth in less than five minutes. The hole, with its tiny Schwarzschild radius, would pull virtually none of the earth's mass into itself during this time. Instead, its gravitational force would constrict a tube of magma as it goes by, drawing inward the material closest to it. But again, this tube of matter would not have time to enter the black hole before it passes by. The tube would be a cylinder of super-compressed liquid rock. After the black hole passes, the tube would rebound, sending out a shock wave through the earth that might well create earthquakes. The matter that had been in the compressed tube would remix with the surrounding magma. The earth's surface would experience the same hammering when the black hole leaves as it did when it entered.

The entire earth would also experience a small dislocation from its normal orbit around the sun first toward Diablo as it approached the earth and then, minutes later in the opposite direction, as the black hole departed. Quake zones worldwide would probably be active for some months following the encounter, leading to much destruction and suffering for the inhabitants of those areas. While we normally think of quakes as occurring along coastlines, such as in

California, Washington, South Carolina, Massachusetts, and New Hampshire, inland faults would also be disturbed, including those in western Nevada, northwestern Utah, eastern Idaho, southwestern Montana, the region where Kentucky, Indiana, Illinois, Missouri, Tennessee, and Arkansas meet, northern New York, and northern Maine.

The black hole would be moving at hypersonic speeds as it traverses the earth's atmosphere. The air pulled in toward it as it passes would be compressed and then would re-expand sharply, creating a sonic boom. The passage would probably cause the air to develop tornadoes.

Effects on the Moon

The moon, too, would be affected by Diablo's passage. Recall that the black hole entered the earth on the side opposite the moon. Upon leaving the earth, the hole would head up toward the moon. Unlike the earth, with its greater mass, the moon would offer much less resistance to being broken apart. In all likelihood, Diablo's tidal effect would tear the moon asunder as the hole passed through it.

The high speed of the black hole would prevent most pieces of the moon from following it. Indeed, the process of disassembling the moon would not even be complete before the black hole leaves the vicinity of the earth and moon. The pieces of the moon would initially drift apart, but then their mutual gravitational attraction would pull many of them back together. The impacts of the moon's rubble recoalescing would have two important effects. First, some pieces would actually be knocked up off the reassembling body and sent earthward. These would strike the earth from time to time, causing damage commensurate with their sizes and speeds.

Second, impacts of the moon's remnants on its new and growing surface would make the new moon molten. It would glow red in our sky for years. Powerful volcanoes on its surface would emit lava earthward into space. Some of this material, too, might reach the earth's surface. If, as seems likely, the remade moon's orbit around the earth is different from its original path, the tides on the earth, the cycle of lunar phases, and the occurrences of both lunar and solar eclipses would also be forever changed.

One thing that becomes clear from this scenario is that it is the "normal" gravitational effects from the black hole, not the more exotic effects of general relativity, that affect the earth and moon. This would be true for the passage near or through the earth of any black hole that doesn't destroy the planet outright.

Effects of the Black Hole's Passage on Life

Denizens of the earth living within a few miles of where the black hole enters would die within minutes. Buildings in nearby cities would buckle and crumble, the ground would sway like Jell-O, and lava would spew a sea of liquid over everything for miles around. The change of ocean tides due to the passage would then set off tidal waves, which would inundate shoreline habitats some hours later. Cities on the shore would also face walls of water as the oceans swayed and surged, adjusting themselves to the changes in the earth's orbit around the sun and the moon's orbit around the earth. These motions of the oceans would be similar to seiches, which are waves that occur in some lakes and seas today caused by water sloshing around. You create seiches when you walk around holding a large pan filled with water, for example.

The emission of gases, especially carbon dioxide, from the volcanoes activated by the passage would cause global increases in temperature over the following months due to the greenhouse effect. The extra carbon dioxide would eventually be absorbed by the oceans and by green plants. Although it would change our weather, it is unlikely that the added carbon dioxide or other volcanic gases would emerge in sufficient volume to suffocate life here.

The Black Hole After Impact

We choose to have the solar system capture the primordial black hole on its way out, as often happens to comets entering the inner solar system for the first time. This is done by having Diablo give up some of its energy to Jupiter. The giant planet, with 318 times as much mass as the earth, would speed up only a little as a result of the black hole passing by. But the black hole would slow down and be pulled into permanent orbit around the sun.

The black hole, like the comets trapped in the inner solar system, would be in a highly elliptical orbit around the sun. Also like many comets, it would periodically cross the earth's path, although the chances that the hole would strike this planet again are very small. The black hole would usually pass above or below the plane of the ecliptic when it is at the same distance as the earth is from the sun. (The same also applies to comets today.) The hole could be located by its gravitational effects on planets and moons. The amount these bodies would deviate from their usual orbits due to the hole's gravitational force as it passed near them would then enable astronomers to calculate its orbit around the sun.

The presence of a black hole in the solar system would be a boon to astrophysicists trying to understand the nature of matter and the universe. By studying it, they would be able to test the details of theories such as Einstein's theory of general relativity, the Hawking process, and the Big Bang theory of the formation of the universe. Furthermore, the energy of the black hole could actually be harnessed to provide electricity.

The Hawking process could be verified by seeing how many particles spontaneously appear from space around the black hole. General relativity's predictions, including the presence and properties of a region just outside the Schwarzschild radius called the ergoregion, could also be checked.

Photons and other objects falling straight inward would actually follow a spiral path, being dragged around by the black hole's rotation. This dragging would be most pronounced in the ergoregion. Matter and light can escape from the black hole's gravitational attraction after entering the ergoregion provided they have enough speed, unlike their fate after passing inside the Schwarzschild radius. Indeed, put into the right orbit, a particle can actually leave the ergoregion with more energy than it had upon entering. Which brings us to how humans could harness the black hole.

Removing Energy from the Black Hole

In 1969, the brilliant British mathematician Roger Penrose proposed a method of removing energy from a rotating black hole. Spacecraft could be sent down into the black hole's ergoregion along paths directly above the black hole's equator and in the same direction that the hole rotates. Keeping in mind that our black hole's ergoregion extends out from the hole's center only a few thousandths of an inch, we are discussing extremely miniature spaceships. But in this day when we can manufacture objects much smaller than a thousandth of an inch, this is not an insurmountable problem.

At a critical time, a jet of particles would be fired from the tiny ship at very high speed into the black hole. If the infalling matter is sent on the appropriate trajectory across the tiny Schwarzschild radius of the black hole, the hole's ergoregion will respond by flinging the remaining part of the craft back outward with even more energy (i.e., faster) than it came in. The spacecraft's extra energy would come from inside the black hole. Unlike the result in the Hawking process, the black hole would not lose mass. Rather, it would give up angular momentum; it would rotate more slowly each time it kicked a spacecraft away.

The disposable, outflying craft would then be captured in a glorified flywheel. A flow of such spacecraft would turn the flywheel, thereby powering an electrical generator. That energy could be stored for use by people living in space or returned to the earth. While the black hole would be too small to provide unlimited energy to the human race, it could provide an enormous amount, helping to make a permanent human presence in space economical and reducing our need to pollute the earth in order to generate energy.

SEEING THE WORLD THROUGH INFRAROSE-COLORED GLASSES: EARTH

WE HAVE EXPLORED SEVERAL ALTERNATIVE SOLAR SYSTEMS, AS well as how the present one would be affected by outside influences. In this chapter we will start applying the "what if" concept to the world as it actually is. Keeping the earth and universe exactly as they are, we will imagine looking at them through new eyes.

The information we receive from our eyes is so important that several large regions of our brains are devoted to collecting and analyzing the data contained in visible light as well as supporting the functioning of the eye itself. Our eyes enable us to perceive objects from inches to countless trillions of miles away. It is only because we have eyes that we know about the global structure of the earth. They also act as our window to the universe beyond the earth's atmosphere. Until this century, everything we knew about the sun, the planets, the moons, the comets, the stars, and the other galaxies came through our eyes. But do they show us everything there is to "see"?

Because eyes provide us with so many details about the

world, we rarely ever wonder if there is knowledge in the parts of the electromagnetic spectrum to which our eyes are not sensitive (namely, radio waves, infrared radiation, ultraviolet radiation, X rays, and gamma rays). In this chapter we will explore that question by asking the following: What would we learn about the earth and space if we could see electromagnetic radiation other than visible light? Along with this, we tackle the issue of whether nature could have evolved eyes sensitive to these other parts of the spectrum.

This chapter's theme originated in the ideas of chapter 5 on the development of our eyes to help us react to dangers and opportunities. However strong our need for electromagnetic information, it was only because of a fortunate set of circumstances that our eyes were to become useful and biologically possible. Consider the sequence of events that must occur before eyes of any kind can be useful. The photons originating in the sun have to travel 93 million miles to reach the earth. Only a small fraction of the sun's radiation flows in our direction. The sun must therefore emit enormous quantities of any kind of photon so that enough of them pass our way to be detected.

To get here, those photons must travel across interplanetary space without being absorbed or scattered by gas or dust particles. Otherwise they would be dispersed and we would see the sun at those wavelengths as if through interplanetary clouds. The photons then have to pass through our atmosphere in quantities sufficient to irradiate the earth's surface. Some of the incident photons must then scatter off the objects on the earth so that eyes can detect them. All these steps so far are independent of whether nature has the biological versatility to evolve eyes sensitive to visible light. That is a separate issue.

Clearly, each of these conditions for having useful eyes is met by light in the visible part of the spectrum. Why was this part of the electromagnetic spectrum chosen? What about the other types of photons: radio waves, infrared, ultraviolet, X rays, and gamma rays? Does the same sequence of events hold for them? If so, what would we see if our eyes were sensitive to these other types of photons? To understand how photons reach our eyes after scattering off objects on Earth, it will help us greatly if we add a few more lines to the tapestry of electromagnetic theory first woven in chapter 5.

THE INTERACTION BETWEEN MATTER AND ELECTROMAGNETIC RADIATION

As we discussed in chapter 5, photons are created in stars. They also come from other sources including fires, light bulbs, some chemical reactions, and radioactive materials. Under the right circumstances, photons are absorbed by electrons orbiting in atoms they encounter. However, only photons with certain energies (or, equivalently, wavelengths) are absorbed by each type of atom. Photons with all other wavelengths pass through the atom unimpeded. Upon being absorbed by an electron, the photon vanishes. To understand this curious property of requiring the correct energy to be absorbed, we consider how photons are different in their interactions with matter than, say, a meteoroid en route to the earth.

Unlike the property of photons just described, every meteoroid on a collision course with Earth strikes the planet or burns up in its atmosphere. There are no records of incoming space debris having passed unscathed through the earth. Photons encountering electrons in orbit, however, do that all the time. This difference in how photons and rocks interact with other bodies is an important feature of quantum mechanics, among the most original physics ever developed. Photons and meteoroids are not the only objects that respond differently to the matter they encounter. In particular, electrons in orbit around atomic nuclei behave very differently from planets orbiting the sun.

Unlike planets, which can orbit at any distance from the sun, electrons can only exist in certain orbits (called "allowed" orbits) around their nuclei of protons and neutrons. Each allowed electron orbit has a well-defined energy, and each different allowed orbit occupies a different region of space around the nucleus. According to the well-tested laws of quantum mechanics, an electron can absorb only photons with energies approximately equal to the difference between the energy of the photon's present orbit and the energy of a higher-level orbit. Most photons not possessing the right energy to loft an electron to another allowed orbit pass through the atom unscathed.

Upon absorbing a photon, each electron moves to an allowed higher-energy orbit, often called an excited state. There it typically stays for less than a millionth of a second. The excited electron then spontaneously loses energy and returns to a lower-energy orbit. Sometimes the electron moves straight back down to the orbit from which it began; sometimes it pauses at one or more intermediate allowed orbits. The term *down* is literal in two ways. First, it indicates a loss of energy, and second, the electron also moves closer to the nucleus with each transition to a lower-energy orbit.

Each transition down is accompanied by the emission of a photon by the electron. The photon has energy (and corresponding wavelength) equal to the difference between the higher energy of the orbit the electron leaves and the lower energy of the orbit in which it lands. One of the important observations about emitted photons is that they are not necessarily emitted in the same direction as the photon that struck the electron in the first place; rather, photons are emitted in all directions.

Every different type of atom and molecule has a unique set of allowed orbits for its electrons. As a result, each type of atom and molecule absorbs and emits a unique set of photons with different energies. This is why objects made of different chemicals have different colors. An object that appears blue emits primarily photons that our brains interpret as blue, while a yellow object emits photons we interpret as yellow. Most objects do not appear as one of the colors of the rainbow because just about everything emits more than one wavelength of visible light. This mixing of primary (rainbow) colors creates all the hues that make the world so vibrant.

The photons that are not immediately absorbed by electrons near the surface penetrate deeper and deeper into the object they encounter. Eventually they either pass through it, strike an electron that is not in orbit around a nucleus, or strike an electron in orbit. The energy of the photons goes into heating the object. Eventually this heat is radiated away as infrared photons, even though the incoming photon might have been visible or ultraviolet. In other words, photons of all kinds can heat objects. With this knowledge and the material in chapter 5 as background, let us begin by seeing why nature evolved eyes sensitive to visible wavelengths.

OUR PRESENT EYES

The intensity of the sun's electromagnetic emissions peaks in the visible part of the spectrum, meaning that the sun sends out more visible photons than any other kind. Since large numbers of photons are needed to illuminate objects at the distance of the earth from the sun, this range of wavelengths has an initial advantage over all other types of photons in being able to illuminate objects. The space between the sun and the earth is virtually devoid of gas and dust, enabling all the visible-light photons heading our way to reach the earth's upper atmosphere without being scattered.

Upon entering the atmosphere, visible-light photons

encounter the nitrogen, oxygen, and water vapor that comprise most of Earth's air. Happily, these gases have few energy levels between electron orbits that correspond to absorbing visible photons. The visible photons that are scattered most by molecules in the air are those we perceive as violet. Scattering is successively less for blue, green, yellow, and orange, with red photons scattered least of all. (This might seem to suggest that the sky should appear violet, rather than blue. All things being equal, that would be true. However, the sun emits fewer violet photons than blue photons, so that even though a larger fraction of the violet photons are scattered, there are fewer of them. And our eyes are less sensitive to violet than to blue, which increases our *perception* that the sky is blue.) So, most visible-light photons zip straight through our atmosphere without colliding with any molecules.

Because relatively few visible photons are scattered in the air, astronomers say that the earth's atmosphere has a window through which this electromagnetic radiation passes. There are also windows for some wavelengths of radio waves, fewer wavelengths of infrared radiation, and fewer still of ultraviolet radiation. X rays and gamma rays have no windows through our atmosphere.

Upon reaching the earth's surface, visible-light photons encounter the matter that exists here. Here, too, visible light behaves in a useful way. In the first place, it does not physically harm anything, as X rays do. Furthermore, most objects on Earth scatter visible light; only those objects that appear jet-black do not. Different objects have different colors because the chemistry of each object determines which color photons it scatters. And so, all the nonbiological aspects of illuminating the earth with visible light photons are favorable.

Finally, nature had to evolve eyes sensitive to visible-light photons. These photons have to be able to stimulate molecules in the eye so that their presence can be detected. This transformation of electromagnetic energy into chemical energy is crucial, since it is the interface between the sense organ and the outside world. Eyes of different species employ several different light-sensitive compounds that are connected to the animal's brain by optic nerves. The light-sensing molecules all work in a similar way: An incoming photon releases a molecule that is weakly attached to a protein called opsin. (The most common light-sensitive compound in the human eyes is rhodopsin.) This chemical separation, stimulated by a photon, is detected chemically and reported to the brain.

An individual photon affects an individual rod or cone (light sensor) in the eye. Photons scattered from adjacent parts of a single object are focused by an eye's lens onto adjacent rods or cones in the back of the eye. Being able to determine where individual photons came from (in the study of optics this is called resolution) enables detailed images of the world around us to be momentarily imprinted on the light-sensitive rods and cones in our eyes. Through millions of rods and cones detecting individual photons, useful images of objects are mapped onto the eye's retina and from there into the brain.

Importantly, visible-light photons are small enough so that many can be detected simultaneously in eyes of all sizes. All visible-light photons are smaller than a ten-thousandth of an inch. Human eyes have roughly 100 million rods and cones.

We just noted that the focusing of photons onto the rods and cones requires a lens through which light can pass. Without adequate focusing, objects do not appear distinct, as anyone who wears glasses knows. Therefore, nature had to evolve a transparent lens as well as a clear coating to protect the eye from the outside world.

Clearly, there are several places in this long, convoluted

chain of circumstances where a physical obstacle could have prevented eyes sensitive to visible-light photons from evolving. For example, these photons could have carried so much energy as to destroy living tissue, or there might have been no transparent molecular compounds for nature to fashion into lenses. Bearing such possibilities in mind, we now examine why our eyes do not see any more of the electromagnetic spectrum than they do. While our limited range of colors certainly gives us adequate information about the world around us, we will find that there are things happening on Earth, and even more importantly, in space, that we cannot see.

GAMMA-RAY- AND X-RAY-SENSITIVE EYES

For a variety of reasons, eyes sensitive to gamma rays and X rays can be easily eliminated as evolutionary possibilities. In the first place, the sun emits billions upon billions of times fewer X rays and gamma rays than it does visible-light photons. These emissions of high-energy photons are also very sporadic. For both of these reasons, many fewer X-ray and gamma-ray photons reach the earth's atmosphere than do visible-light photons.

While X rays and gamma rays do travel through interplanetary space, none of them reach the earth's surface. Instead, they are all absorbed by molecules in the upper atmosphere. This is just as well because upon impact, X rays and gamma rays tear apart the molecules they strike, often freeing electrons as well. If such activity occurred frequently on Earth, the resulting destruction of biological molecules would sterilize the planet.

Focusing X rays and gamma rays is also no mean feat. Unlike visible light, they do not change direction when going through lenses. Rather they plow into the lenses, destroying them. We will return to this issue when we discuss how astronomers have developed telescopes to detect X rays and gamma rays from astronomical sources.

X-ray photons are emitted from certain naturally occurring radioactive elements on Earth. Their presence poses health risks to nearby animals and plants, but such substances are rare enough that it was not necessary for creatures to evolve X-ray sensors of any kind. Radioactive elements are those that spontaneously transform or decay from one element into another. The decay process is often accompanied by the emission of X rays. These X rays usually travel only a few yards before they are absorbed by molecules in the atmosphere or by objects on the surface. Among others, natural isotopes of uranium, potassium, rubidium, radon, and carbon are radioactive.

Perhaps the most significant natural X-ray hazard on Earth is radon. Radon is a colorless, odorless, tasteless gas that seeps out of rocks such as granite. When this gas concentrates, as it does inside houses, the X rays it emits do pose health risks, including lung cancer. It is worth noting that the normal exposure to medical and dental X rays poses little hazard because the doses are so low and the exposure times so short. The body easily repairs damage done by these X rays.

Not knowing what they would find, astronomers began searching for astronomical X-ray sources in brief rocket flights during the late 1940s. These were followed by a series of small, orbiting X-ray satellites in the early 1970s. Similar to the problems experienced by the eyes of early animals, early X-ray telescope "eyes" were crude and insensitive. They saw only the strongest X-ray sources, and even then had trouble locating them precisely. The first gamma-ray telescopes were put in orbit in the late 1960s. Like their X-ray counterparts, they "saw" gamma rays in space indistinctly.

Just as visible-light eyes evolved and improved in sensitivity and resolving power over millions of years, so too have X-ray and gamma-ray telescope "eyes" evolved. These radiations are focused by having them skim along the surfaces of tube-shaped telescopes, changing direction slightly each time they hit the tube.

Astronomical X-Ray Sources

By far the most fascinating X-ray images taken so far are of the sun. Whereas in visible-light photographs the sun appears as a uniform disk occasionally peppered with sunspots, in X-ray light it looks heavily mottled, with some regions emitting large quantities of X rays, and others virtually none at all. Indeed, in many X-ray photographs the sun does not even appear round.

Some two thousand other X-ray sources are known. Among these are black holes, hot stars, supernova remnants, pulsars, the centers of some galaxies, and groups of stars called globular clusters.

Supernova remnants (see chapter 6) are particularly interesting X-ray sources. The radiation has two sources. First is the collision between the expanding remnant and the pre-existing interstellar gas and dust (collectively called the interstellar medium). Similar to the infalling matter striking the accretion disk of a black hole, the rapidly expanding shell of gas and dust of the supernova rams into the interstellar medium, thereby generating and emitting X rays. The remnant loses energy as it interacts with the interstellar medium. As a result the remnant slows down and eventually becomes part of the interstellar medium. By that time the X ray and other emissions associated with the remnant cease.

The second source of X rays occurs in those supernova remnants in which the remnant star left behind is a rotating neutron star with a powerful magnetic field that does not pass through the star's rotation axis. Such remnant bodies, rotating anywhere from once every few seconds to a thousand times a second, are called pulsars. Rotating stars with off-axis magnetic fields are believed to be the rule, rather than the exception. (This applies even for stars on the Main Sequence. Indeed, our sun rotates about once every twenty-seven days, and its magnetic field does not emerge along the axis of its rotation.)

As the neutron star rotates, its magnetic field is dragged around with it. When the changing field encounters charged particles from the remnant, such as electrons or protons, the field causes the particles to accelerate. Such motion causes the particles to emit X-ray and/or other electromagnetic radiation. Therefore, both pulsars and the supernova remnants around them often emit X rays as a result of the pulsar's motion.

Astronomical Gamma-Ray Sources

There have been fewer gamma-ray sources detected in space than any other type. This reflects the fact that of all types of electromagnetic radiation, gamma rays are the hardest to generate and detect. Most gamma-ray sources that have been detected so far appear to originate in the Milky Way. The very heart of our galaxy, called the nucleus, emits both X rays and gamma rays. Recall that as seen from Earth, this region is in the direction of the constellation Sagittarius.

The Milky Way's nucleus cannot be seen through visible-light telescopes because the intervening gas and dust clouds scatter all the outbound visible-light photons. Therefore, through high-magnification visible-light telescopes, Sagittarius appears filled with interstellar clouds: Some look white; some are colorful. Everything we know about the galaxy's nucleus comes from the nonvisible radiation it emits, especially radio waves, infrared radiation, and X rays.

Some observations suggest that the source of gamma rays at the heart of the galaxy is a million-solar-mass black hole, but the precise mass and even the black hole's very existence are still controversial. (Other observations suggest the presence of a black hole in the nucleus containing only one hundred solar masses.)

Gamma-ray sources also include the sun, interstellar gas, supernova remnants, and neutron stars. Some neutron stars appear to be associated with intense bursts of both gamma rays and X rays (other than the pulsar mechanism). These objects are called bursters, and the details of their behavior are still being derived. Suffice it to say that when a gamma-ray burster erupts, it is brighter through a gamma-ray telescope than even the (much, much closer) sun as seen through the same telescope.

RADIO-SENSITIVE EYES

The evolution of radio-sensitive eyes meets with obstacles at several crucial steps. We begin at the source, with microwaves, the sun's most intense radio emissions. The sun emits millions of times fewer microwaves than it does visible light. And the sun's weakest radio photons are millions of times weaker than even its microwave output.

Virtually all radio photons pass straight through interplanetary space. And like visible light, radio waves have a broad window through the earth's atmosphere to the ground: All radio photons with wavelengths shorter than about thirty feet and longer than an eighth of an inch pass through the air without being absorbed or reflected.

A second problem for radio eyes occurs when radio waves encounter matter on the earth's surface. Being so much larger than individual atoms, most radio photons pass through matter rather than being absorbed by it and scattering off it like visible light. We experience this effect daily inasmuch as we can listen to radios even inside windowless rooms. The radio waves pass, albeit attenuated, through glass, wood, brick, concrete, and innumerable other materials. They scatter well only off objects rich in metals, especially those containing copper, aluminum, and other good conductors of electricity.

To compensate for the low scattering rate of radio photons, radio-sensitive eyes would have to be larger than visible-light eyes just to collect more of the few available radio photons. There is, in principle, one way around this problem that would not involve larger eyes: Have the eyes look at each object for a longer period of time. By staring at something for a longer period, small radio eyes could collect as many photons as do larger ones. In practice, however, this would not be useful to animals, since as we have already discussed, eyes serve as distant early-warning devices. Eyes are designed for supplying information on dangers and opportunities that typically occur in matters of seconds. Having to look at something for minutes or hours in order to see it would defeat the purpose of natural eyes.

The increase in size needed for eyes to capture photons rapidly enough to see objects would be compounded by the separate need to see details. We discussed earlier how our present eyes see whole images because individual photons strike separate rods or cones. The same would have to apply to radio eyes; namely, individual radio photons should hit individual radio rods and cones. However, instead of being a few millionths of an inch across like visible-light photons, radio photons are all longer than a thirtieth of an inch. Therefore, individual radio rods and cones would have to be at least a thirtieth of an inch apart. Eyes with such sensors would have to be tens of thousands of times larger than their visible-light counterparts.

Given these two considerations, the total area of radio eyes would have to be millions of times larger than the area of visible-light eyes. Assuming that the biological technology of radio rods and cones would not be insurmountable for nature, a useful radio eye would be a ludicrous three hundred feet across. Carrying around a pair of those would make it very hard indeed to get close to someone else. So, for several reasons, radio eyes would not serve the evolutionary purpose of helping us see and respond to the world around us.

Astronomical Radio Sources

The tremendous difficulties associated with making detailed observations of radio emissions belie the awesome variety of radio-emitting objects in space. The first radio telescope was developed in 1931 by Karl Jansky of Bell Telephone Laboratory in New Jersey. Jansky's discovery of astronomical radio sources was a combination of serendipity and persistence. As with so many scientific discoveries, Jansky was not looking for what he found. His job was to track down sources of radio noise that might interfere with radiotelephones, then under development. He detected a source of static that he could not connect with anything on Earth. Carefully noting the time of day when it was active, he determined that the noise source was out in space, outside the solar system. Radio astronomy had begun.

Between then and today, radio-telescope "eyes" have evolved more rapidly than technological eyes in any other part of the electromagnetic spectrum. By coupling together several telescopes, sometimes separated by thousands of miles, radio-telescope eyes can now see more detail from radio-emitting objects than do telescopes in any other part of the spectrum, including the very best visible-light telescopes. *Also, radio telescopes can see some objects that are completely invisible to optical telescopes.* Some otherwise invisible radio-emitting objects are as small as the earth, while others are much larger than the largest visible galaxies in the universe.

Among the innumerable radio sources in space are the center of the Milky Way galaxy, interstellar and intergalactic gas clouds, intergalactic gas jets, pulsars, and, in the solar system, the sun and Jupiter. We turn now to a brief discussion of some of these objects as seen through human-made radio-telescope eyes.

Astronomers estimate that between 5 and 10 percent of the matter in the disk of the Milky Way galaxy is in the form of interstellar gas and dust. Much of this material is located in clouds that float between stars and stay together because of their own gravitational force. Most of this gas and dust cannot be seen in visible-light telescopes. We know of its presence and location primarily by its radio emissions.

Many galaxies emit streams or jets of gas that have overflowed the visible bounds of the galaxy and spread into intergalactic space. As this out-flowing gas loses speed, it forms intergalactic clouds so enormous as to fill millions of times more space than the visible galaxies from which it originated. Consider, for example, the intriguing gas jets associated with the visible galaxy Centaurus A. In visible light this galaxy appears round (we think it is spherical). Also visible are broad streams of gas and dust bisecting the galaxy. In a visible-light photograph, the galaxy appears to be bursting apart or perhaps parts of it are colliding with one another. In any event, radio images of Centaurus A reveal two otherwise-invisible jets of gas extending outward from this visible image and perpendicularly to the visible gas and dust. The radio jets extend much farther outward than the diameter of the visible galaxy, ending in two huge intergalactic gas clouds. The jets and clouds give astronomers important clues about the tumultuous activity inside this galaxy.

Turning to the solar system, radio emissions from Jupiter teach us many things about that planet that we couldn't learn from the sunlight it reflects. The bulk of Jupiter is a sea of liquid hydrogen and helium. It may have a core of rock and metal, but that material is all thousands of miles below the planet's surface. Jupiter is completely surrounded by a permanent cloud cover. While we can't see its surface from space, radio emissions from the vicinity of Jupiter's liquid surface pass right through the clouds into space.

Some of Jupiter's radio emissions reach the earth. From their intensity and wavelengths we can determine the rate at which the body of the planet rotates (once every nine hours, fifty-five minutes, and thirty seconds). This rotation rate is about five minutes faster than the clouds orbiting the planet's equator. Furthermore, from Jupiter's radio emissions with wavelengths between ten to a hundred inches long, we have learned that there is an invisible oval cloud of gas, very similar to our Van Allen belts, surrounding the planet. Because of this cloud's orientation, we also know that like the earth's, Jupiter's magnetic fields do not leave the planet's surface along its axis of rotation. All this and much more is known because radio telescopes enable us to "see" things invisible to our eyes.

Finally, there is the intriguing issue of "seeing" the surface of Venus. Venus is enshrouded with a permanent mantle of clouds. The only way visible images of the surface can be taken is by parachuting a spacecraft through the cloud layer to the clear sky below it and then transmitting the pictures back to Earth. As Soviet space scientists learned with their Venera series of landers between 1967 and 1978, landing on Venus poses two extremely difficult problems. First, the clouds are filled with sulfuric acid, which rapidly degrades the equipment on spacecraft falling through them. Second, as we discussed earlier, the pressure at the surface of Venus created by its carbon dioxide atmosphere is over one hundred times greater than the pressure we feel from the air on the surface of the earth. At the same time, the surface temperature is over nine hundred degrees Fahrenheit. It is extremely difficult to build spacecraft to withstand such pressures and temperatures for more than a few hours.

All the Soviet Venera landers that were functioning when they reached the surface survived for less than an hour. Like the proverbial blind person meeting the elephant, planetary astronomers tried to describe the surface of Venus based on a handful of photographs of the local terrain taken by two landers. What the first photos showed were plains strewn with rocks. Some rocks looked smooth, as if weathered by the heavy atmosphere. At the other site rocks looked rougher, suggesting that they were either young or that the air had worn them down less. No hint was given of the diverse terrain and structures later seen by the radio-wave maps of the planet's entire surface.

The alternative to landing and taking visible-light photographs is to make pictures of the surface using radio waves. As on Jupiter, radio waves can pass through the clouds of Venus without scattering. Therefore, if the surface of Venus emitted sufficient quantities of radio photons, the entire planet could be "observed" from an orbiting spacecraft equipped with a radio eye (that is, a radio antenna and receiver). Such an observation system would not have to land, so it would not have to withstand the extreme temperature and pressure of Venus.

Unfortunately, Venus is not a strong radio emitter in its own right, and the sun does not emit enough radio photons to cause the planet's surface to shine radio light. To see the surface in reflected radio waves, astronomers had to create their own radio "flashlight" to illuminate the surface so that it can be seen by the spacecraft's radio eye. This technology, first developed for use by satellites making images of the earth from orbit, exists and has been used to map the entire surface of Venus. The resulting maps show features as small as a football field there (although no football teams were found).

There are direct equivalents between the radio eyes used in orbit around Venus and biological eyes sensitive to visible light. The antenna on the spacecraft serves as the radio eye's lens. The receiver at which the radio photons are focused does the same thing as rods and cones. The transmission of data to Earth works like an optic nerve. The data-analysis computers back on Earth do what the visual cortex in the brain does, namely, making usable images out of the raw data transmitted to them.

While the techniques used by computer to interpret radio waves are somewhat more complicated than those used by our brains to make sense of what our eyes see, the result of each is an image that we can sensibly interpret. As a result of this radar imaging, we now have more global knowledge about the surface of Venus than we do about some of the ocean bottoms on the earth.

ULTRAVIOLET-SENSITIVE EYES

The intensity of ultraviolet radiation from the sun falls off precipitously relative to the visible-light levels it emits. The ultraviolet photons with wavelengths just shorter than violet light are the most copious. These are called near-ultraviolet photons, as compared to the more energetic but less numerous far-ultraviolet photons. We discussed in chapter 5 how the ozone in the earth's atmosphere prevents most of the sun's near-ultraviolet radiation from reaching the earth. However, some of these lower-energy ultraviolet photons do have windows to the earth's surface. They cause both suntans and sunburns.

Furthermore, in large numbers these photons scatter off a limited number of objects on Earth. Ultraviolet-sensitive eyes would have a limited but detectable supply of photons.

In determining the minimum size ultraviolet-sensitive eyes could be, the lower number of photons compared with visible light would somewhat offset the fact that ultraviolet photons are all shorter than visible light. Consequently, ultraviolet-sensitive eyes would not have to be much larger than visible-light eyes.

We do not see the relatively well-scattered near-ultraviolet photons primarily because this radiation is absorbed in the lenses of our eyes. In other words, the lenses of most animals are opaque to ultraviolet. People who have artificial lenses made of ultraviolet-passing plastics can see some near-ultraviolet radiation scattered off objects on Earth. Some insects, such as honeybees, have eyes naturally sensitive to near-ultraviolet photons. Because near-ultraviolet radiation can pass through clouds, bees can see many flower petals even during overcast times, when the flowers do not reflect much visible light. This ability gives them an edge in finding food over other animals without ultraviolet-sensitive eyes.

But possessing eyes tuned to near-ultraviolet radiation carries considerable cost, which is more than most animals can afford. Bees cannot see red, orange, and yellow light. These colors are emitted more intensely by the sun than ultraviolet and scattered more intensely by many more objects on Earth. While seeing near-ultraviolet radiation scattered from their food source (flowers) is a clear-cut advantage for bees, it is more useful to most animals to have eyes sensitive to longer-wavelength visible light rather than shorter-wavelength ultraviolet radiation.

Astronomical Ultraviolet Sources

The development of ultraviolet-sensitive telescope eyes for astronomical exploration has proven very useful. Most ultraviolet observations are done in space, since the atmosphere excludes most ultraviolet photons. Orbiting ultraviolet telescopes include the Hubble Space Telescope and the International Ultraviolet Explorer. Ultraviolet radiation is detected from the atmospheres of hot stars and from interstellar clouds. Because the composition of stars' atmospheres gives us valuable information about their internal activities, these observations are important to astronomers.

Finally we explore the infrared part of the spectrum. These photons, with wavelengths just longer than the visible color red, provide more information about the surface of the earth than does the rest of the nonvisible spectrum. We will see that even where nature has not exploited that potential to the fullest, humans have begun to do so.

INFRARED-SENSITIVE EYES

The sun gives off a wide range of infrared wavelengths, the most intense of which are almost as bright as the red light it emits. Infrared radiation passes unscattered through interplanetary space. In fact, infrared is less affected by the occasional gas and dust in the solar system than is visible light, which tends to scatter easily. However, the lowest-energy infrared photons are scattered directly back into space by the earth's atmosphere. The intermediate-energy infrared photons are absorbed by carbon dioxide and water molecules in the air. The energy absorbed from these photons makes the gas molecules vibrate or move, thereby warming the air. Finally, a broad range of the highest-energy infrared photons reaches the ground intact.

Most materials on Earth scatter some of the infrared photons that strike them. Just as with the scattering of visible-light photons, the intensities and wavelengths of scattered infrared photons vary from material to material, depending on an object's chemical composition and the texture of its surface. But there is a major difference between the emission of visible photons and that of infrared photons by matter here. Virtually all visible-light photons emitted by objects on Earth come from sources such as the sun, fires, and human-made lights. However, only a fraction of the infrared photons each object emits is due to scattering of incoming photons. The rest are internal photons previously stored in the object.

As discussed previously, not all of a body's stored heat arrives as infrared photons. Much of it comes as visible light that is absorbed (instead of scattered). Photons that are absorbed in matter cause atoms in the object they strike to vibrate. This vibration is microscopic, of course, and the motion manifests itself as heat. In this way, the energy in visible photons is transformed into heat energy. Each object can contain only a limited amount of heat, depending primarily on its chemical composition and mass. Upon reaching this limit, every object begins re-emitting as much energy as it receives. The object is then said to be in thermal equilibrium. However, the energy reradiated does not go out as the same types of photons that struck in the first place. Rather, the emitted photons are mostly infrared (heat).

When the sun goes down, an object that was in thermal equilibrium during daylight hours continues to radiate its stored heat as infrared photons. Without their being replenished, the object cools down. Thus, *matter gives off infrared radiation even when it is not being illuminated by the sun.* This suggests that infrared-sensitive eyes could be very useful to nocturnal animals.

Warm-blooded animals have another major source of heat besides the sun: their bodies. The internal temperature of warm-blooded animals is kept constant, usually well above that of the surrounding air and land. Some of this self-generated heat is emitted as infrared radiation through the skin and out the lungs with respired air. When warm-blooded creatures are hotter than their surroundings, they emit more infrared photons than do cold-blooded creatures or inanimate objects (by Wien's displacement law; see chapter 5). Therefore, warm-blooded animals are very bright infrared emitters compared with cold-blooded creatures or most inanimate objects. Warm-blooded creatures would stand out like beacons at night when seen through infrared-sensitive eyes.

Nocturnal animals equipped with such eyes would have a distinct advantage over creatures without them, since they would be able to see their foes and prey in pitch-black or stormy conditions. Snakes, who are nocturnal hunters, already have rudimentary but serviceable infrared-sensitive eyes. Lying in wait, they can detect their warm-blooded prey by the heat they emit. Snakes are cold-blooded. Even if their quarry also had infrared-sensitive eyes, it would have a very hard time seeing the snake against the background rock and sand, which would be emitting the same intensity of infrared radiation as the snake itself.

Despite the fact that few animals have infrared-sensitive eyes, the usefulness of infrared radiation to all living things has not been lost on nature. Along with most other animals, we humans have primitive heat-detecting nerves on our skin. That, of course, is how we can feel heat from an object without having to touch it. We even have enough heat-sensitive nerves to determine the general direction of heat sources. The trouble is that we don't have enough of these nerves to obtain detailed images of the heat-emitting objects. Being able to detect heat from a distance helps prevent animals from burning themselves.

With the wealth of information that infrared radiation can provide, especially at night, it would seem likely that more sophisticated infrared-sensitive eyes would have evolved. The fact that they did not suggests that there are important design difficulties inherent in putting a lot of infrared detectors together in eyes.

One important problem is that warm-blooded bodies

have internal temperatures of around one hundred degrees Fahrenheit and thus would give off the same radiation that the infrared-sensitive eyes would be seeking from the outside world. If infrared sensors were packed together at the backs of eyes, like our rods and cones, the heat generated by the body surrounding them would go through the walls of the eyes and set off the sensors. This internally generated background glow would severely limit the eye's ability to separate out the photons it receives externally from those created inside the viewer's body. While this obstacle is not insurmountable, it might have been more of a challenge to nature than it was worth for the benefits gained.

There are at least two biological solutions for the internal heat problem. First, nature might evolve a biological insulation and refrigeration scheme that would keep the infrared eyes cooler than the rest of the body. In that way, since the infrared rods and cones would not receive as many photons from inside the body, they would be more sensitive to photons from outside. An alternative is to separate the infrared eyes from the rest of the body by putting them on stalks. This would enable the air to keep them cool. Such eyes could be enhanced by an aerodynamic design that funnels cooling air past the hottest parts of the eye.

Both of these approaches are very costly in an evolutionary sense. Cooling fluid would have to be prechilled before circulating around the eyes. This could be done by bringing the fluid to the body's surface to remove its heat to the air. Then the cooled fluid would have to travel through insulated vessels to the eyes. Otherwise, the body's heat would rewarm it. The plumbing would be quite complex, not to mention that in tropical climates the surface cooling would work poorly, if at all. On the other hand, eyes on stalks would be very susceptible to damage from blows and from wind. It would also be harder to keep them aimed in the right direction in order to get good stereo (three-dimensional) images. Visible-light eyes today do not have either of these last problems because they are firmly supported inside rigid skulls.

How large would natural infrared eyes have to be? We know that the wavelengths of infrared photons are just slightly longer than the wavelengths of visual-light photons. Also, we know that the sun emits near-infrared photons almost as copiously as it emits visible light. Therefore, we surmise that suitably cooled infrared eyes need only be a few times larger than our present eyes to see details of infrared-emitting objects.

The evolutionary value of seeing objects on the earth through infrared eyes is great enough that we humans have already developed extensive night-vision technology. The original purpose for this equipment was to enable military forces to see at night, thereby enabling them to fight and defend themselves around the clock. This was demonstrated vividly in the Persian Gulf war. The forces of the Gulf Coast Allies, using night-vision equipment, kept up a twenty-four-hour air barrage against the Iraqis. It is clear that an army equipped with such night-vision gear has an overwhelming advantage over otherwise equal foes who lack this apparatus.

Infrared-imaging technology is now available to civilians. The entire world was enthralled by the night photography we saw of antiaircraft fire over Baghdad taken with infrared-sensitive cameras. We actually saw more of the city through these infrared images than did any gunners involved in the battle who did not have infrared equipment. Here at home, infrared photographs are used to show where heat is leaking from houses. This helps people determine where their houses need insulation to help cut heating bills.

Also, infrared-sensitive eyes would give humans a much better sense of how people around them are feeling. The amount of heat our bodies give off depends on what we are thinking and feeling. For example, when people become angry, blood rushes to their skin. They then radiate more of the heat stored in the blood, and if seen in infrared light, they would appear brighter than normal. Similarly, when people become sexually aroused, their skin often becomes flushed with blood and they literally become hotter (and brighter) as seen in the infrared. Conversely, when people become scared, blood leaves the surface of their skin and they become darker in the infrared.

Astronomical Infrared Sources

Infrared astronomy is done at high altitudes because water vapor in the air absorbs so many of the infrared photons. Specifically, infrared observations are done high on mountaintops, from high-flying airplanes, or from satellites orbiting the earth.

There are many astronomical objects that can be seen from the infrared radiation they emit. The moon is visible in the infrared because of the sunlight it absorbs and re-emits. The sunlit part of the moon visible from Earth has a temperature of 265 degrees Fahrenheit, compared to the unlit (dark) side, which is at −170 degrees. Because of this temperature difference, each part of the moon, sunlit and dark, has a different infrared brightness. Indeed, the moon goes through the same cycle of phases as seen in infrared that it does in visible light.

Jupiter is another bright source of infrared radiation. The giant planet re-emits sunlight in the infrared part of the spectrum as well as giving off heat that it generates internally. Indeed, Jupiter emits more than twice the heat it receives from the sun. Seen through an infrared-sensitive telescope, Jupiter has the same striped appearance of belts and zones that it has through a visible-light telescope. This indicates that different regions of Jupiter's upper atmosphere, which is what we see, have different temperatures. Likewise, the other giant planets Saturn, Uranus, and Neptune give off measurable infrared.

There are events in our solar system during which infrared radiation but no visible light is emitted. These include asteroids hitting comets and asteroids colliding with each other. If we could see them, these impacts would appear as bright bursts of infrared light because of the heat they generate. After each collision the heated debris cools and darkens. Similarly, meteoroids that hit the moon generate significant amounts of infrared radiation (and some visible light). These impacts are often powerful enough to melt rock and create craters.

Our view through infrared eyes of objects outside the solar system would be equally interesting. Consider first the space between the stars. These regions are filled with thin, irregular mixtures of gas and dust. The dust absorbs visible light passing through it and reradiates it as infrared. There are also thick clouds of gas and dust inside of which new stars are forming. Visible light from these new stars does not reach us because it is absorbed in the dust clouds surrounding them. However, we would be able to see the clouds themselves glowing brightly in infrared that results from their being heated by the stars inside them. Indeed, infrared telescopes do see such hot clouds.

The best location to see a cloud of interstellar matter is in the middle of the constellation Orion's sword. If you look at the middle "star" in Orion's sword with a pair of binoculars or a small telescope, you will see that it isn't a star at all. Rather, it is a gigantic gas and dust complex called the Orion molecular cloud, part of an even larger cloud region that covers the entire image of Orion. This area is a hotbed of contemporary star formation. While the newborn stars are invisible inside their cocoons, the surrounding gas and dust glow brightly as seen in the infrared. Stars begin to shine in visible light only after they disperse their gas and dust coverings.

Like the sun, all other stars emit infrared as well as visible light. Infrared eyes would see the stars as well as the clouds. But many of the stars would not appear as points of light, as they do to our present eyes. Besides the dense molecular clouds throughout the Milky Way galaxy, there are thin interstellar filaments of cirruslike clouds that scatter infrared radiation from the stars. This scattering adds a general haze to the infrared light emitted by individual stars and clouds. This emission, along with the infrared from the gas and dust clouds dispersed around young stars, makes many stars appear as smudges, rather than points, as seen from their infrared emission. Infrared light spread by glowing dust creates an Impressionist heaven—an infrared van Gogh sky.

There are fascinating and unexpected astronomical activities to be seen in all parts of the electromagnetic spectrum. But as far as nature is concerned, the evolution on Earth of eyes sensitive to nonvisible photons has been limited to a few species that can detect narrow bands of ultraviolet or infrared radiation. For most animals there is just too little useful information that non-visible radiation can supply to justify the tremendous evolutionary effort that would have been needed to develop alternative eyes for all creatures.

FROM NEW WORLDS TO OUR WORLD: WHAT IF THE OZONE LAYER WERE DEPLETED? EARTH

IMPORTANT INSIGHTS CAN BE GAINED BY ASKING "WHAT IF" QUES-
tions about any aspect of our world; indeed, people use
"what if" questions to explore the world all the time. Scientists often begin studying new facets of nature by asking
questions like, "What would happen to the fish if this fertil-
izer gets into the rivers?" or "What will happen to the particle production rate if I increase the energy of the particles
in the synchrotron beam?" Lawmakers try to forecast the
results of their legislative actions by including their new
laws in "what if" scenarios and then studying the changed
world they have created. The armed forces have elaborate
computer programs to help them study "what if" scenarios
on the battlefield. In fact, although you may never be aware
of it, you instinctively create "what if" worlds every day. If
you want to hear it done consciously, listen to a six-year-old
child some day. If my oldest son is any example, young
children often spend active days creating and exploring
new worlds by asking, "What if . . . ?"

The "what if" process is an essential part of our ability

to consider the long-term effects of our actions before we take them. Whereas other animals act upon immediate, physical urges, people usually take time to consider the implications of their actions first. "What if I marry that person? What if I take that job? What if I move?" The insights gained are important in our decision making.

When we consider "what if" questions, our thoughts alternately focus on narrow issues and then range broadly among apparently unrelated concepts. We begin by framing the question at hand, which contracts our thoughts into a few words defining a specific issue, which we then explore in detail. Thoughts crisscross our minds, making new connections, raising new questions, developing new scenarios. This often leads to the need for further, more broad-ranging thought. The more often we repeat this cycle, the further into the new situation we go and the greater insight we gain from it.

Suppose I wanted to buy a new car. Thoughts about its safety, cost, size, comfort, and style first drift through my mind. I focus on the issue of safety: "What if the car is hit from the front? How likely is it that my passengers and I will survive the crash? How expensive would repairs be? Does it have air bags? What if the car is hit from the side? Will the doors buckle and crush me?" I consider each of these questions separately. Then I turn to the car's cost: "What can I save if I get a car with a four-cylinder, rather than a six-cylinder, engine? What would it cost if I got antilock brakes?" Since this latter question reminds me of safety again, I need to weigh the importance of this feature relative to its cost. The process continues. As you can see, asking "What if . . . ?" is clearly a part of decision making in each of our lives.

We attempt to bridge the gap between the world-defining and the everyday uses of "what if" questions by examining a human-made problem that is affecting the entire earth: the destruction of the ozone in the earth's upper atmosphere. By asking the specific question "What will happen to the earth and life on it if this process continues until a quarter of the ozone is gone?," we will bring together some of the science introduced throughout this book. We will see, as was argued in the introduction, that we humans are no longer just travelers on this planet.

STRATOSPHERIC OZONE DEPLETION

This book began with a discussion of the moon's role in the evolution of the earth and its atmosphere. We explained how ultraviolet radiation from the sun helped start evolution by allowing inorganic chemicals to combine on the early earth. In chapter 5 we pointed out that the advent of molecular oxygen in the earth's atmosphere allowed ozone to form high up in the stratosphere. The ozone proved especially effective in absorbing ultraviolet radiation, thereby preventing it from reaching the earth's surface. For over 350 million years the ozone layer has provided protection from lethal ultraviolet radiation. This protection was essential for life to move onto Earth's landmasses.

However, in the past fifty years the ozone layer has begun to dissipate. As we will see, stratospheric ozone depletion affects other changes already taking place in the atmosphere as a result of human activities. These include increases in carbon dioxide and other greenhouse gases (with the resulting change in global air temperature), the creation of smog (changes in lower air chemistry), and even the development of a second, lower-level ozone concentration.

Note that ultraviolet radiation is separated into three wavelength ranges. Normally, only some of the longest-wavelength, lowest-energy ultraviolet radiation reaches the earth's surface; the rest of this lowest-energy ultraviolet is absorbed by ozone. The penetration through the atmosphere of the lowest-energy ultraviolet photons became an asset to life when nature harnessed these photons to help form the D vitamins. These steroidlike compounds are involved in regulating the body's use of calcium, which is crucial for building bone. Virtually all the intermediate-wavelength ultraviolet photons are absorbed by ozone. Ozone is not as important in screening out the shortest-wavelength, highest-energy ultraviolet radiation because normal molecular oxygen helps absorb much of it.

Causes of Ozone Depletion

The natural ozone layer that has protected life on the earth's surface from ultraviolet radiation for hundreds of millions of years is located in the stratosphere between nine and thirty miles above the earth's surface. The ozone is mixed with other gases, primarily nitrogen and molecular oxygen; it does not exist in its own, separate layer. As mentioned earlier, there is now a secondary ozone layer directly over our heads in the troposphere. While the tropospheric ozone will come into the discussion later, we will now focus primarily on the stratospheric ozone layer, referring to it, for convenience, as the ozone layer.

Recall that ozone forms when the sun's ultraviolet radiation breaks molecular oxygen into two atoms of oxygen. One or both of these atoms can combine with another oxygen molecule to form ozone. Alternatively, one or both of these atoms can combine with ozone, converting the ozone to molecular oxygen. Since there is so much less ozone in the stratosphere than there is molecular oxygen, this latter process happens relatively infrequently.

Ozone is not created at a uniform rate because its formation depends on the amount of ultraviolet radiation emitted by the sun, which varies. The most noticeable variation in solar radiation is over an eleven-year cycle that coincides with the number and location of sunspots.

Ozone is depleted by a variety of natural and human-made gases. Natural gases consuming it include, among others, methyl chloride from oceans, forest fires, and volcanoes; sulfuric acid from volcanoes; water vapor; and finally, nitrogen compounds. Human-made substances that damage the ozone layer include, among others, compounds containing chlorine, bromine, nitrogen, and oxygen. There used to be an equilibrium between the rate at which ozone was destroyed and the rate at which it formed. With the ascent of human-made gases into the stratosphere, that balance no longer exists. Human-made chemicals traveling up to the stratosphere are now destroying ozone faster than the sun can create it.

Some gases, such as methane, can enhance either the formation *or* the depletion of ozone, depending on the latitude and altitude. Methane helps create ozone in the lower stratosphere, but high over the south polar regions of Earth it has the opposite effect by aiding the formation of polar stratospheric clouds, which destroy ozone. Methane is emitted naturally from chemical decomposition in swamps, rice paddies, and animal intestines. However, humans have greatly increased the amount of methane by raising more cows (renowned for methane production) and growing more rice than would otherwise be here, and by cutting down forests, which provides food for vast numbers of methane-generating termites, among other things.

The most important chemical element involved in ozone depletion today is chlorine (individual bromine atoms actually deplete ozone more effectively, but fewer bromine-containing gases are presently drifting up toward the stratosphere). The human-made chlorine that reaches the stratosphere today has been released into the air as a variety of manufactured compounds. Over 80 percent of the human-made compounds implicated in destroying ozone by supplying chlorine to the stratosphere are members of the family of chlorofluorocarbons, or CFC's. Over 15 percent of the destruction comes from carbon tetrachloride and methyl chloroform. All these compounds are able to rise into the stratosphere because they are very stable; there is no energy source in the lower atmosphere (the troposphere) to break them down before they reach the stratosphere. These compounds then float up through the stratosphere until, reaching its upper levels, they encounter enough ultraviolet radiation from the sun to break them down, freeing the chlorine in them to act on the ozone.

CFC's and Stratospheric Ozone Depletion

CFC's are molecules composed of chlorine, fluorine, carbon, and sometimes other elements. They were first created in 1928 for use in the refrigeration industry as nonflammable, relatively nontoxic replacements for the flammable, toxic ammonia and sulfur dioxide that were used as refrigerants at that time. To the joy of the chemical industry, CFC's also proved to be extremely inert, meaning that they did not interact with many other substances. As a result, engineers found many other uses for CFC's: in air conditioning, as propellants in innumerable spray cans, as gases to clean surfaces of debris, and as gases to help blow styrofoam into molds.

The injection of CFC's into the earth's atmosphere is also an excellent example of how environmental-impact studies often fall short of their mark. Early studies on the impact that CFC's have on the atmosphere were limited solely to the effects on the troposphere. Because it was found that CFC's were inert in this region, it was concluded that they would not affect the air at all. No one bothered to consider the effects of CFC's in the stratosphere until 1973, when two chemists, Mario Molina and F. Sherwood (Sherry) Rowland, stumbled across the fact that CFC's profoundly affect the ozone layer.

Molina had recently received his doctorate and had begun a postdoctoral research position with Rowland, professor of chemistry at the University of California at Irvine. Molina was studying the effects of chlorine on ozone in the earth's stratosphere. In December 1973 he discovered and Rowland confirmed that if chlorine somehow got into the stratosphere, it would transform large quantities of ozone there into molecular oxygen.

Several things conspired to make their work more than an academic exercise. Rowland and Molina discovered that once established in the stratosphere, a single atom of chlorine would transform tens or even hundreds of thousands of ozone atoms into molecular oxygen before being washed back down to Earth. It would not take very much chlorine to seriously deplete the incredibly thin, diffuse ozone layer. There is so little ozone surrounding the earth and protecting us from ultraviolet radiation that if all of it were compressed to the density of the air we breathe, it would make a layer less than one-eighth of an inch thick!

Therefore, if enough chlorine were injected into the stratosphere, it could destroy the majority of the ozone much more rapidly than the sun could replace it. To their horror, the two chemists discovered that CFC's were in fact being made in sufficient quantity to cause the serious damage they had predicted. Even after they and others notified the scientific and political communities, CFC production continued to increase. After all, several large companies were heavily invested in CFC production, and even more companies were involved in its use. They all resisted change, demanding more and more proof that CFC's were truly causing damage to the ozone layer. By 1985, over 1 billion pounds of CFC's were being manufactured each year.

As early as 1982, British researchers discovered what they believed to be a decrease in the ozone layer over

Antarctica. The results were confirmed and made public by 1984. Even that discovery was not sufficient to force a halt in CFC production. Many believed that the notorious ozone hole could have been caused by other, possibly natural, factors.

Conclusive proof of the complicity of CFC's in damaging the ozone layer came in 1987. In 1986 and 1987, researchers traveled to Antarctica and to the southern tip of South America in concerted efforts to learn the cause of the "ozone hole" that was forming over the South Pole each September and October. There were several competing theories. Besides the chlorine reactions, other causes, such as nitrogen reactions, solar activity, and natural air flows (dynamical theories) had been proposed. By 1987 all of the theories except the chlorine theory had been eliminated: The smoking gun for chlorine-mediated depletion of the Antarctic ozone came from measurements of chlorine in the stratosphere. The measurements led to two curves; one showed the ozone dropping, while the other showed the simultaneous increase in chlorine concentration (actually chlorine monoxide, an intermediate compound in the chain of events that removes ozone).

While nature inserts limited amounts of chlorine into the air in volcanic eruptions, most of it up there is human-made. Chlorine is depleting the ozone layer in two general ways. All around the world chlorine is converting ozone to molecular oxygen faster than the sun can replace it. Yet, ever since the stratospheric chlorine levels became appreciable, this global loss of ozone has still been quite small, probably amounting to less than 1 or 2 percent of the total ozone layer.

The other source of ozone loss, taking place at the South Pole, is decidedly more extensive and, therefore, more sinister. It occurs because some of the chlorine in the stratosphere drifts toward the South Pole. There, during

August and September, the chlorine becomes trapped and concentrated by a natural whirlpool called the Antarctic Vortex. This high density of chlorine then attacks the ozone over Antarctica, creating the hole. Keep in mind, as we discussed in chapter 4 on Urania, that the earth's axis is tilted relative to the plane of its orbit around the sun. Therefore Antarctica goes through a period of several months when the sun is continuously below the horizon.

The ozone hole comes into being just when the sun begins rising over Antarctica in September. By late October the continent begins warming and the Antarctic Vortex dissipates, freeing the chlorine to spread back over the rest of the earth. By this time the ozone over Antarctica has been depleted by as much as 95 percent. The ozone builds back up over Antarctica, but—and this is the kicker—most of the replacement ozone is created not by the sun's radiation breaking up molecular oxygen but from existing ozone drifting southward from other parts of the stratosphere. In other words, each year some of the ozone that is over your head goes south to plug the renewed Antarctic ozone hole.

This seemingly local Southern Hemisphere disaster is rapidly pumping down the *global* ozone concentration. If the ozone hole didn't occur and if all CFC production stopped immediately, it is estimated that global ozone depletion would peak at losses around 10 percent lower than normal. But because of the ozone hole and the slow cessation of CFC production, the amount of depletion could go to 30 percent or higher.

Atmospheric scientists have discovered some of the basic chemistry relating to the depletion of ozone. There is still an enormous amount to be learned because of the complexity of the atmosphere and of the interactions taking place within it. With this as background, we choose to consider the effects of a 25 percent global ozone loss, which is well within the realm of possibility.

Effects of Increased Ultraviolet Radiation

The first major effect of ozone depletion is that more ultraviolet radiation reaches the earth's surface. Ultraviolet radiation is divided by wavelength into three ranges. From longest (least hazardous) to shortest wavelength, these are denoted UV-A, UV-B, and UV-C rays. With the ozone layer at its normal level (keep in mind that even before human intervention the amount of ozone in the atmosphere varied considerably in cycles ranging from daily to over several years), most of the ultraviolet radiation getting to the earth's surface was UV-A, with only limited amounts of UV-B and virtually no UV-C radiation getting through at all.

As the ozone layer wanes, more UV-B radiation is able to penetrate into the troposphere and to the earth's surface. Perhaps the only good news is that UV-C is blocked by molecular oxygen, so that even with a calamitous 25 percent drop in stratospheric ozone, the deadly UV-C photons would still be kept from the earth's surface.

In the Oceans: Case Study of a Disrupted Food Chain

We consider first some of the changes in aquatic food chains created by increased ultraviolet radiation entering the oceans of the world. At the very base of the ocean food hierarchy are a variety of plant life collectively called phytoplankton. These tiny life forms scavenge raw molecules such as phosphates, nitrates, and silicates from the ocean, turning them into organic compounds and using sunlight for energy. Phytoplankton are then eaten by zooplankton and krill. These latter two groups are small sea animals that exist in vast numbers and are consumed, in turn, by many kinds of birds, seals, whales, and other large sea creatures throughout the oceans of the world.

Because they rely on sunlight to generate their food, phytoplankton live close enough to the ocean's surface to absorb sufficient numbers of photons to function. This upper region of the ocean where light penetrates is called the marine euphotic zone. The depth at which visible light ceases to penetrate, called the euphotic depth, varies from a few feet to several hundred feet, depending on the clarity of the water. Ultraviolet radiation penetrates the top 10 percent of the euphotic zone, where phytoplankton receive the most sunlight and are most productive in duplicating. Many types of phytoplankton, including the ubiquitous diatoms, are very sensitive to ultraviolet radiation. Staying at their present depths, many of these life forms could not survive the large increase in ultraviolet radiation that would accompany the drop in ozone.

The increased ultraviolet radiation would prevent many types of phytoplankton from living and reproducing in their normal habitat in the upper reaches of the euphotic zone. Such plants would have to live at greater depths, where the ultraviolet radiation is less harmful, or they would have to very quickly evolve resistance to ultraviolet radiation to avoid becoming extinct.

The zooplankton that now feed on ultraviolet-sensitive phytoplankton would find their food source abruptly missing as the radiation level rises. They would have to find new food at the same level, descend to greater depths to follow their food source, or become extinct. Descending means adapting to colder temperatures, greater pressures, and less light. Moving to different strata also implies competing with other creatures already living at those levels.

On the other hand, those phytoplankton that are able to quickly evolve to withstand the increased ultraviolet might also change in ways that would alter their suitability as food sources for the animals that ate them in their previous incarnations. Zooplankton unable to eat their traditional food would have to find new sources of nourishment or they, too, would become extinct. All this just goes to show

how complex matters can become even at the very bottom of the food chain as ultraviolet levels increase in the oceans.

Fish, birds, and sea mammals that feed on zooplankton would find their favorite food sources missing or inedible. Many diving birds would especially suffer in this scenario, since they cannot descend as deeply or as long as fish can to locate their traditional food sources at greater depths. (The increased pressure that accompanies greater depth would greatly stress their relatively fragile frames.) Numerous species of diving birds would therefore face extinction. As the numbers and distributions of these larger species decline, the carnivorous fish and mammals at the top of the food chain would also be affected.

Some species of fish are more susceptible than others to cancers and other diseases caused by exposure to increased ultraviolet radiation. As a result of both direct and indirect challenges to them, the total and relative numbers of sea creatures will change. This redistribution of biomass (matter in living organisms) would eventually affect virtually every life form in the ocean, even those living below the euphotic zone. Many of these latter life forms are scavengers. The changes in the types and numbers of dead life forms descending to them would alter their eating and breeding patterns.

At the very top of the ocean food chain, humans would see changes in the number, types, and locations of food fish. Some species of fish now consumed by humans or used in our industries would be less available, changing the eating habits of vast numbers of people. Today 30 percent of the animal protein that humans consume comes from the sea. The changes in available fish and sea mammals would in all probability cause severe famine for some time.

Some species abundant in the ocean food chains now may not be so after the transition. Some species would remain plentiful, but would become part of different food chains than they are presently in. That is, they would have different prey and different predators with which to contend. While some species would almost assuredly become extinct, the oceans themselves would probably not become lifeless, liquid deserts. The ability of most species to evolve, even within a few generations if necessary, in response to physical and biological changes in their environment, suggests otherwise.

On the Land

Increased ultraviolet radiation would harm life on land as well. As ultraviolet radiation penetrates the skin of a person, animal, or plant, the photons encounter biological molecules, some of which they alter by ripping apart molecular bonds. Living organisms have evolved the ability to repair or replace damaged molecular structures *up to a point,* beyond which they cannot keep up with the destruction. Mutations leading to cancers and other diseases then result.

We consider three separate areas of impact on humans of the increased ultraviolet radiation: effects on the skin, effects on the internal organs, and effects on the eyes. Even when the ozone layer is at a "normal" concentration, the ultraviolet radiation reaching the earth's surface causes suntans, sunburns, and with sufficient overexposure, skin cancer. The increase in intermediate-energy (UV-B) ultraviolet photons because of the decreased ozone layer means that more ultraviolet would be penetrating deep into exposed skin. This is especially true in people with light skin, since melanin, the dark skin pigment, is an exceptionally good absorber of ultraviolet radiation. People with lighter skin and less melanin are therefore more at risk of contracting ultraviolet-mediated diseases, such as skin cancer.

There are three major types of skin cancer caused by ultraviolet radiation: basal cell carcinoma, squamous cell carcinoma, and different types of melanomas. Since the basal and squamous cells are screened by melanin, dark races have a much lower incidence of cancers of these cells today.

Assuming people don't modify their sunbathing and sun-exposure habits, the number of new basal and squamous cell cancers that would occur with a stable 25 percent drop in the ozone layer has been estimated at over a billion in a fifty-year period. Treatment for both basal and squamous cell carcinomas is very effective if done early. Even so, with the large number of new cases, millions of deaths can be expected from these cancers, especially in regions of the world where medical assistance is less than adequate.

Melanomas are extremely dangerous and hard-to-treat cancers. Death rates even today are around 40 percent within five years of contracting such a cancer. The protracted period of intense UV-B radiation would make it even harder to control them. It is likely that the vast majority of the melanomas caused by the increased ultraviolet radiation would be fatal. Epidemiologists estimate that if ozone levels dropped by one quarter, there would be over 2.5 million deaths caused by malignant melanoma over a fifty-year period.

The second group of problems caused by ultraviolet radiation occurs below the skin. Like AIDS, ultraviolet radiation suppresses human autoimmune activity. This means that ultraviolet radiation decreases the body's ability to detect and fight disease. Suppressed immune capability would lead to a dramatic rise in a number of diseases that a body with a normally functioning autoimmune system can usually prevent, such as pneumonia, leprosy, and internal cancers. Therefore, there would be a dramatic jump in death rates from diseases caused indirectly by ultraviolet radiation.

It is instructive to note that the devastation caused by

the three types of skin cancer and the internal diseases would be suffered primarily by light-colored races. Therefore, over a long-enough period of excess ultraviolet exposure, the distributions of races on Earth would measurably begin to change.

We turn now to increased ultraviolet effects on eyes. Snow blindness (acute or photo-keratitis) is caused by the efficient reflection of ultraviolet light by snow. These ultraviolet photons irritate eyes, often causing them to swell shut. This is an extremely painful but usually reversible condition. An increase in ultraviolet radiation would lead to more "snow blindness," even in climates without snow.

More debilitating would be the increased number of cataracts caused by the excess ultraviolet radiation. Cataracts are often caused by ultraviolet radiation today, too, but the number of cases would increase by an estimated 100 million or more over a fifty-year period with the decrease in the ozone layer.

Increased ultraviolet radiation would require people to protect themselves when they go outdoors by using sunscreen, wearing clothes that prevent penetration of ultraviolet to the skin, wearing hats and, very importantly, ultraviolet-blocking sunglasses. Indoor protection would also be needed because ultraviolet radiation passes through glass normally used in windows. The easiest way to keep ultraviolet out of buildings would be to treat all windows with so-called low-e coatings that prevent the radiation from penetrating the glass.

Animals would not be immune to diseases caused by ultraviolet radiation. Some members of most species would face blindness from cataracts. Species without sufficient fur or other protection would also contract skin diseases. The fact that different species have different sensitivities to ultraviolet radiation implies that food chains on land, like those in the sea, would be disrupted by the drop in atmospheric ozone.

Different types of plant life also have different immunities to increased UV-B radiation. As a result, stable communities of vegetation would be disrupted as once-dominant plant species fail and less ultraviolet-sensitive competitors take their places. Pine trees, for example, are known to suffer when ultraviolet radiation is increased; these would decline in number, their space being filled with heartier vegetation.

Excessive ultraviolet radiation affects plant growth in several ways. Generally, stem and leaf growth decreases, photosynthesis slows, and for many varieties, the dry weight of full-grown plants decreases. Therefore, cultivated plants that are sensitive to ultraviolet radiation, such as peas, beans, melons, cabbage, soybeans, and mustard, among others, would have much lower yields than they do now. This would add to the strain of feeding the world population.

Ultraviolet radiation today is notorious for fading carpets and furniture. These problems would be greatly aggravated when the ozone level drops. Items made of rubber and plastic would tend to degrade or weaken more rapidly than they do today. Outdoor products using these materials would become embrittled more quickly by the increased concentration of ultraviolet photons. Paints and wood stains would fade more rapidly. In general, exterior surfaces would require more maintenance than they do today.

In the Atmosphere

The air we breathe is greatly affected by the ultraviolet radiation passing through it. Consider first its composition. The air here at the bottom of the troposphere is composed of 78 percent nitrogen molecules, 21 percent oxygen molecules, nearly 1 percent argon atoms (an inert gas), and traces of over one hundred other compounds. Most of these trace gases occur naturally as well as from human activity.

For example, carbon monoxide occurs as a result of forest fires and through the oxidation of methane, while humans create it by burning hydrocarbons and coal. Nitrous oxides (compounds of nitrogen and oxygen) occur naturally during forest fires, electrical storms, and anaerobic activity in soil, while humans create these compounds by burning oil, gas, and coal. Sulphur dioxide (which smells of rotten eggs) is emitted into the air during volcanic activity and in the oxidation of hydrogen sulfide, while humans create it in the combustion of oil and coal and in processing sulphur ores. There are many other examples.

There are also chemical compounds in the air exclusively as a result of human activity. These include complex aromatic hydrocarbons created in burning gasoline and in the evaporation of gasoline, paint, and solvents; chloroform, used in medicine as well as in solvents, gasoline combustion, and the bleaching process for wood pulp; and the chlorofluorocarbons, as discussed earlier.

Overall, trace-element levels in the air have more than doubled as a result of human activity. The increase in numbers and types of molecules has led to important changes in the overall chemistry of the air we breathe. One of the most important changes is the dramatic rise in greenhouse gases, especially carbon dioxide, water vapor, methane, and the CFC's, among others.

The increase in greenhouse gases leads to an increase in the temperature in the atmosphere and hence on the earth's surface. The effects of greenhouse warming on the earth would easily fill a separate chapter. In order to keep focused on the effects of increased ultraviolet radiation from ozone depletion, we will not explore the implications of greenhouse warming for the earth, but recall that many of its effects are discussed in the context of other imagined worlds throughout this book.

Consider the effects of ultraviolet radiation on air.

Among the trace elements in the troposphere, several absorb ultraviolet radiation even today. This absorption of ultraviolet photons leads to chemical changes in these gases. Therefore, as the ozone layer becomes more and more depleted in our scenario, the increased concentration of ultraviolet photons passing through the troposphere would cause an increase in the chemical activity of the lower atmosphere. Based on what is happening today, we can extrapolate on how the atmosphere would change and how those changes would affect life.

The changes in the chemistry of gases under the influence of electromagnetic radiation such as ultraviolet light is called photolysis. The most significant photolysis in the lower atmosphere is the creation of ozone from a variety of sources, most notably the nitrogen-oxygen compounds. Other compounds created directly as a result of photolysis include carbon monoxide, hydroxyl radicals (unions of one oxygen and one hydrogen atom), nitrous oxide (laughing gas), and hydrogen. These molecules are highly reactive, creating other compounds in turn. There are over a thousand chemical reactions taking place among trace gases in the atmosphere, many requiring ultraviolet radiation at one step or another.

The ozone created in the troposphere is a major component of smog. Therefore smog—which is already horrendous in and around such cities as Los Angeles, Denver, Tokyo, London, and Mexico City, to name a few—would become even more noxious as a result of the increase in ultraviolet radiation passing through the troposphere.

Tropospheric ozone is a mixed blessing. On the one hand, an increase in this gas would actually help to prevent some of the ultraviolet radiation that got through the stratospheric ozone layer from reaching the earth's surface. However, so much of that ozone would be at ground level that it would have its own injurious effects on all forms of life. Ozone is a very active molecule. It reduces the ability of plants to photosynthesize. It inhibits their flowering and germination. It even helps enhance microbial and insect infestations in conifers and possibly other types of trees. As a result, we can expect a massive decline in vegetation, especially around smog-covered cities. Some areas that are undergoing deforestation today, such as the Black Forest of Germany, would probably be completely decimated by the ozone and the acid rain (specifically sulfuric acid) that ozone helps form in the atmosphere. Less of the carbon dioxide emitted by humans and our activities would be used in photosynthesis because of the tremendous decrease in plant life. The carbon dioxide content of the atmosphere would therefore rise, adding to the increase in greenhouse warming.

Ozone is also dangerous for humans and animals to breathe, causing and exacerbating a variety of respiratory illnesses as well as irritating eyes. Many people today who suffer from respiratory ailments find days with high ozone levels uncomfortable. With an increase in tropospheric ozone, the concentration of surface ozone and the number of days each year when the ozone level is high would both rise dramatically. This would lead to hundreds of thousands more deaths and millions more cases of respiratory and eye problems each year than our polluted air causes today. Similar conclusions apply to the effects of the carbon monoxide created by ultraviolet photons, but on a smaller scale. All in all, the air will be much more dangerous for all life on Earth.

Effect of Ozone Depletion on the Weather
With unprecedented deforestation, food-chain disruptions, millions more premature deaths, and tens of millions more diseases each year caused directly or indirectly by the increase in ultraviolet radiation passing through the ozone

layer, what more could go wrong? The answer is: the weather. Today the weather results from a combination of horizontal and vertical motions of the air, combined with collection and deposition of moisture and dust from the surface of the earth. The horizontal motion of the air is caused by the earth's rotation dragging the air around with it as well as changes in air pressure in different places around the earth. Air flows into regions where the pressure is low and out of regions where the pressure is high. The vertical motion of the air is caused primarily by heating from the earth below. Upon being warmed, the air expands and rises, cooling as it goes up. (A secondary cause of vertical motion is the deflection of horizontal winds by mountains and valleys.)

The stratospheric ozone is important in confining the vertical motion of the air, which in turn affects the horizontal motion, as we will see shortly. To understand the ozone's role in our weather, consider the rise of air from the earth's surface. As anyone who has ever climbed or driven up a mountain knows, the air gets cooler at higher altitudes. Expanding, cooling air rising from the earth's surface continues upward until it has released enough heat to cool to the temperature of the air around it. The cooled air then moves horizontally until it loses enough heat to become even cooler than the surrounding air. It then descends back toward the earth's surface. Upon reaching the earth's surface again, the air moves horizontally until heated enough to rise once more. In fact, much of the air moves in rectangular loops composed of horizontal motion along the earth's surface, followed by rising, followed by horizontal motion high above the earth, followed by descending motion.

This cycle of rising and falling occurs all over the troposphere. The maximum height reached by air rising from the earth's surface is called the tropopause, which is, on average, nine miles above the earth's surface. The tropopause is the top of the troposphere and the bottom of the stratosphere.

What makes the troposphere and stratosphere distinct entities is the greater concentration of ozone normally present in the stratosphere than that in the troposphere. Some of the energy in the ultraviolet photons absorbed by ozone molecules in the stratosphere is converted into heat. This heat goes to warming all the stratospheric air. So, while the air temperature falls from the earth's surface upward to the tropopause, *the air, heated by ozone, actually gets warmer as one ascends through the stratosphere.*

Air stops rising at the tropopause because there is warmer air *above* it. The rising, colder air in the troposphere is denser and therefore heavier than the warmer, thinner air at the bottom of the stratosphere. Thinner air rises into denser air, just as a bubble rises in water; denser, cooler air cannot rise into thinner, warmer air. If the ozone layer were depleted by 25 percent, the stratosphere would receive much less heat from the ultraviolet radiation passing through it. As a result, the tropopause would change altitude and become less stable. This means that the rising air in the troposphere would ascend to different levels than it does today before topping out and beginning to move horizontally.

The change in vertical motion brought about by the displaced tropopause would affect the weather on the earth's surface, since the vertical motion of the air is connected to the horizontal motion of the air at the earth's surface. Could the unusual weather being experienced all around the world today be a precursor to the profoundly different weather that would occur if the ozone level plummeted and the tropopause moved? We do know that the ozone layer has already fallen several percentage points from its normal concentration. But attributing the weather change to the present decrease in ozone would require many more observations and calculations than have been done thus far. Given the complexity of the earth's atmosphere, the possibility does remain open.

Unlike the other worlds we have created, the implications of a drastic drop in the concentration of the ozone layer are based on actual observations and computations. This work has been done primarily in light of the current crisis of ozone depletion. If we continue to pump out CFC's and halons (compounds similar to CFC's, but with one or more bromine atoms), the events described in this chapter will almost certainly occur along with others such as the effects of greenhouse warming.

Asking "what if" has helped put the major issues out before us. Can we prevent these problems? Even if we immediately stop their production, most of the CFC's and halons we have already put into the air have not even reached the stratosphere yet. They take anywhere from forty to over a hundred years to ascend to that height. Even if not another drop of ozone-destroying gas is ever emitted, the ozone layer will continue to dissipate for at least the next one hundred years as previously produced CFC's arrive there. By stopping CFC and halon production now, we may keep the total ozone depletion down to 10 percent, rather than 25 percent. All the effects described in this chapter would still happen, but on a smaller scale. Could we restock the ozone layer before things get out of hand? Possibly. But keep in mind how unsuccessful we have been historically when we have meddled in the global activities of nature.

Where do we go from here in the matter of ozone depletion and excess ultraviolet radiation? There is no single or easy answer to that question, but two things are certain: We all have a stake in the result, and we had better act soon.

BIBLIOGRAPHY

In preparing this bibliography, I have grouped books according to the topics for which I used them. As the titles show, many books contain a much wider range of materials than the topics under which they are listed. Those books that are easily accessible to non-professionals are indicated by asterisks.

ASTRONOMICAL DATA

*Bishop, Roy L., ed., *Observer's Handbook 1988*. Toronto: The Royal Astronomical Society of Canada, 1988.

Hopkins, Jeanne, *Glossary of Astronomy and Astrophysics,* 2d. ed. Chicago and London: University of Chicago Press, 1980.

Lang, Kenneth R., *Astrophysical Formulae*, 2d ed. New York: Springer-Verlag, 1980.

GENERAL ASTRONOMY

*Abell, George O., *Exploration of the Universe,* 4th ed. Philadelphia: Saunders Golden Sunburst Series, 1982.

Arnett, W. D., Hansen, C. J., Truran, J. W., and Tsuruta, S., eds., *Cosmogonical Processes*. Utrecht, The Netherlands: VNU Science Press, 1986.

Danby, J. M. A., *Fundamentals of Celestial Mechanics,* 2d ed. Richmond, Va.: Willmann-Bell, 1988.

*Hartmann, William K., *The Cosmic Voyage*, 1992 ed. Belmont, Calif.: Wadsworth Publishing Co., 1992.

*Jastrow, Robert, *Red Giants and White Dwarfs*, 1990 ed. New York and London: W. W. Norton & Co., 1990.

Kutter, G. Siegfried, *The Universe and Life: Origins And Evolution*. Boston and London: Jones and Bartlett, 1987.

*Menzel, Donald H., Whipple, Fred L., and deVaucouleurs, Gerard, *Survey of the Universe*. Englewood Cliffs, N.J.: Prentice-Hall, 1970.

*Michell, John, *A Little History of Astroarchaeology: Stages in the Transformation of a Heresy*. London: Thames & Hudson, 1977.

*Pasachoff, Jay M., *Astronomy: From the Earth to the Universe*, 4th ed. Philadelphia: Saunders College Publishing, 1991.

*Shapiro, Stuart L., and Teukolsky, Saul A., *Black Holes, White Dwarfs, and Neutron Stars*. New York: John Wiley & Sons, 1983.

Wood, John A., *The Solar System*. Englewood Cliffs, N.J.: Prentice-Hall, 1979.

Zeilik, Michael, Gregory, Stephen A., and Smith, Elske V. P., *Introductory Astronomy & Astrophysics*, 3d ed. Fort Worth, Tex.: Saunders College Publishing, HBJ, 1992.

PLANETARY ASTRONOMY

Black, David C., and Matthews, Mildred Shapley, *Protostars and Planets, II*. Tucson, Ariz.: University of Arizona Press, 1985.

*Broecker, Wallace S., *How to Build a Habitable Planet*. Palisades, N.Y.: Eldigio Press, 1985.

*Cole, G. H. A., *Inside a Planet*. Hull, England: Hull University Press, 1986.

Gehrels, Tom, ed., *Protostars and Planets*. Tucson, Ariz.: University of Arizona Press, 1978.

*Kopal, Zdeněk, *The Realm of the Terrestrial Planets*. New York and Toronto: Halsted Press, John Wiley & Sons, 1979.

*Morrison, David, and Samz, Jane, *Voyage to Jupiter*. NASA SP-439. Washington, D.C.: U.S. Government Printing Office, 1980.

Morrison, David, and Wells, William C., eds., *Asteroids: An Exploration Assessment*. NASA Conference Publication 2053. Washington, D.C.: NASA Scientific and Technical Information Office, 1978.

ATMOSPHERES

*Barbato, James P., and Ayer, Elizabeth A., *Atmospheres: A View of the Gaseous Envelopes Surrounding Members of Our Solar System*. New York: Pergamon Press, 1981.

[*]Bohren, Craig F., *Clouds in a Glass of Beer.* New York: Wiley Science Editions, John Wiley & Sons, 1987.

[*]————, *What Light Through Yonder Window Breaks.* New York: John Wiley & Sons, 1991.

Bohren, Craig F., and Fraser, Alistair B., "Colors of the Sky," in *The Physics Teacher,* May 1985, pp. 267–72.

Brown, Robert A., *Fluid Mechanics of the Atmosphere.* San Diego: Academic Press, 1991.

Henderson-Sellers, A., *The Origin and Evolution of Planetary Atmospheres.* Bristol, England: Adam Hilger, 1983.

McEwan, Murray J., and Phillips, Leon F., *Chemistry of the Atmosphere.* New York: John Wiley & Sons (a Halsted Press book), 1975.

[*]Moran, Joseph M., and Morgan, Michael D., *Meteorology: The Atmosphere and the Science of Weather,* 3d ed. New York: Macmillan Publishing, 1991.

Rowland, F. Sherwood, and Isaksen, I. S. A., eds. *The Changing Atmosphere.* New York: John Wiley & Sons, 1988.

Wayne, Richard P., *Chemistry of Atmospheres,* 2d ed. Oxford: Clarendon Press—Oxford, 1991.

Whitten, Robert C., and Prasad, Sheo S., eds., *Ozone in the Free Atmosphere.* New York: Van Nostrand Reinhold, 1985.

BIOLOGY

[*]Alerstam, Thomas, *Bird Migrations.* Cambridge and New York: Cambridge University Press, 1990.

[*]Burton, Robert, *Nature's Night Life.* Poole, Dorset, U.K.: Blandford Press, 1982.

Dejours, Pierre, *Principles of Comparative Respiratory Physiology.* Amsterdam: Elsevier/North Holland Biomedical Press, 1981.

[*]Kappel-Smith, Diana, *Night Life: Nature from Dusk to Dawn.* Boston: Little, Brown & Co., 1990.

Lyman, Charles P., Willis, John S., Malan, André, and Wang, Lawrence C. H., *Hibernation and Torpor in Mammals and Birds.* New York and London: Academic Press, 1982.

[*]Orr, Robert T., *Animals in Migration.* London: Collier-Macmillan, 1970.

Voet, Donald, and Voet, Judith G., *Biochemistry.* New York: John Wiley & Sons, 1990.

BIOLOGICAL CLOCKS

*Moore-Ede, Martin C., Sulzman, Frank M., and Fuller, Charles A., *The Clocks That Time Us: Physiology of the Circadian Timing System.* Cambridge, Mass., and London: Harvard University Press, 1982.

Pengelley, Eric T., *Circannual Clocks: Annual Biological Rhythms.* New York: Academic Press, 1974.

*Winfree, Arthur T., *The Timing of Biological Clocks.* New York: Scientific American Library, 1987.

GENERAL GEOLOGY OF THE EARTH AND MOON

*Bloom, Arthur L., *The Surface of the Earth.* Englewood Cliffs, N.J.: Prentice-Hall, 1969.

Broecker, W. S., and Peng, T. H., *Tracers in the Sea.* Palisades, N.Y.: Eldigio Press, 1982.

Brosche, P., and Sündermann, J., eds. *Earth's Rotation from Eons to Days.* Berlin: Springer-Verlag, 1990.

Cameron, A. G. W., and Benz, W., "The Origin of the Moon and the Single Impact Hypothesis IV," in *Icarus,* vol. 92, pp. 204–216. New York: Academic Press, 1991.

*Carey, S. Warren, *Theories of the Earth and Universe: A History of Dogma in the Earth Sciences.* Stanford, Calif.: Stanford University Press, 1988.

Cazenave, Anne, ed., *Earth Rotation: Solved and Unsolved Problems.* Dordrecht, The Netherlands: D. Reidel Publishing Co., 1986.

*Clark, Sydney P., Jr., *Structure of the Earth.* Englewood Cliffs, N.J.: Prentice-Hall, 1971.

*Cloud, Preston, *Oasis in Space.* New York and London: W. W. Norton & Co., 1988.

*———, *Cosmos, Earth and Man: A Short History of the Universe.* New Haven, Conn.: Yale University Press, 1978.

*Dalrymple, G. Brent, *The Age of the Earth.* Stanford, Calif.: Stanford University Press, 1991.

*Dole, Stephen H., *Habitable Planets for Man.* New York: Blaisdell Publishing Div. of Ginn & Co., 1964.

*Eicher, Don L., *Geologic Time,* 2d ed. Englewood Cliffs, N.J.: Prentice-Hall, 1976.

Eicher, Don L., McAlester, A. Lee, and Rottman, Marcia L., *The History of the Earth's Crust.* Englewood Cliffs, N.J.: Prentice-Hall, 1984.

Fetter, C. W., Jr., *Applied Hydrogeology.* Columbus, Ohio: Charles E. Merrill Publishing, 1980.

*Hartmann, William K., "Birth of the Moon," *Natural History,* November 1989.

*Kandel, Robert S., *Earth and Cosmos.* Oxford and New York: Pergamon Press, 1980.

*Laporte, Leo F., *Ancient Environments,* 2d ed. Englewood Cliffs, N.J.: Prentice-Hall, 1979.

Phinney, Robert A., *The History of the Earth's Crust: A Symposium.* Princeton, N.J.: Princeton University Press, 1968.

*Ringwood, A. E., *Origin of the Earth and Moon.* Berlin: Springer-Verlag, 1979.

*Skinner, Brian J., and Porter, Stephen C., *Physical Geology.* New York: John Wiley & Sons, 1987.

*Strahler, Alan H., and Strahler, Arthur N., *Modern Physical Geography,* 4th ed. New York: John Wiley & Sons, 1992.

EVOLUTION

*Davoust, Emmanuel, *The Cosmic Water Hole.* Cambridge, Mass.: The MIT Press, 1991 (also listed under SETI, since book has two distinct parts).

*Duve, Christian de, *Blueprint for a Cell: The Nature and Origin of Life.* Burlington, N.C.: Neil Patterson Publishers, 1991.

*Dyson, Freeman, *Origins of Life.* Cambridge: Cambridge University Press, 1985.

*Keosian, John, *The Origin of Life,* 2d ed. New York: Reinhold Book Co., 1968.

Mason, Stephen F., *Chemical Evolution: Origin of the Elements, Molecules and Living Systems.* Oxford: Clarendon Press, 1991.

*Osterbrock, Donald E., and Raven, Peter H., eds., *Origins and Extinctions.* New Haven, Conn.: Yale University Press, 1988.

SEARCH FOR EXTRATERRESTRIAL INTELLIGENCE

*Adelman, Saul J., and Adelman, Benjamin, *Bound for the Stars.* Englewood Cliffs, N.J.: Prentice-Hall, 1981.

*Asimov, Isaac, *Extraterrestrial Civilizations.* New York: Crown Publishers, 1979.

*Baugher, Joseph F., *On Civilized Stars: The Search for Intelligent Life in Outer Space.* Englewood Cliffs, N.J.: Prentice-Hall, 1985.

*Davoust, Emmanuel, *The Cosmic Water Hole.* Cambridge, Mass.: The MIT Press, 1991.

*Goldsmith, Donald, and Owen, Tobias, *The Search for Life in the Universe*. Menlo Park, Calif.: Benjamin/Cummings Publishing Co., 1980.

*Hart, Michael H., "Habitable Zones About Main Sequence Stars." *Icarus*, vol. 37, 1979, pp. 351–57.

*Horowitz, Norman, *To Utopia and Back: The Search for Life in the Solar System*. New York: W. H. Freeman & Co., 1986.

ICE AGES

*Asimov, Isaac, *The Ends of the Earth: The Polar Regions of the World*. New York: Truman Talley Books, Dutton, 1975.

*Imbrie, John, and Imbrie, Katherine Palmer, *Ice Ages: Solving the Mystery*. Cambridge, Mass.: Harvard University Press, 1979.

MOON LORE

*Brueton, Diana, *Many Moons*. New York: Prentice-Hall, 1991.

*Guiley, Rosemary Ellen, *Moonscapes: A Celebration of Lunar Astronomy, Magic, Legend, and Lore*. New York: Prentice-Hall, 1991.

*Katzeff, Paul, *Moon Madness and Other Effects of the Full Moon*. Secaucus, N.J.: Citadel Press, 1981.

TIDES

*Clancy, Edward P., *The Tides: Pulse of the Earth*. Garden City, N.Y.: Doubleday & Co., 1968.

*Defant, Albert, *Ebb and Flow: The Tides of Earth, Air, and Water*. Ann Arbor, Mich.: University of Michigan Press, 1958.

Melchiorm, Paul, *The Earth Tides*. Oxford: Pergamon Press, 1966.

Waves, Tides and Shallow-Water Processes. Milton Keynes, England: Pergamon Press in association with The Open University, 1989.

*Wylie, Francis E., *Tides and the Pull of the Moon*. Brattleboro, Vt.: Stephen Greene Press, 1979.

PHYSICS

*Bohren, Craig F., "Understanding Colors in Nature," *Pigment Cell Research*, vol. 1, 1988, pp. 214–22.

*Cutnell, John D., and Johnson, Kenneth W., *Physics*, 2d ed. New York: John Wiley & Sons, 1992.

Jenkins, Francis A., and White, Harvey E., *Fundamentals of Optics*, 4th ed. New York: McGraw-Hill, 1976.

ENVIRONMENT

*El-Sayed, Sayed Z., "Fragile Life Under the Ozone Hole," *Natural History,* October 1988, pp. 73–80.

Elton, Charles S., *The Ecology of Invasions By Animals and Plants.* London: Methuen & Co., 1958.

*Goudie, Andrew, *The Human Impact on the Natural Environment,* 3d ed. Cambridge, Mass.: The MIT Press, 1990.

Groves, R. H., and Burdon, J. J., eds., *Ecology of Biological Invasions,* Cambridge: Cambridge University Press, 1986.

*Jones, Robin Russell, and Wigley, Tom, *Ozone Depletion: Health and Environmental Consequences.* New York: John Wiley & Sons, 1989.

*McKibben, Bill, *The End of Nature.* New York: Anchor Books, 1989.

Milne, Lorus J., and Milne, Margery, *Ecology Out of Joint: New Environments and Why They Happen.* New York: Charles Scribner's Sons, 1977.

*Piel, Gerald, *Only One World: Our Own to Make and to Keep.* New York: W. H. Freeman & Co., 1992.

*Roan, Sharon, *Ozone Crisis: The 15-Year Evolution of a Sudden Global Emergency.* New York: John Wiley & Sons, 1989.

*Shea, Cynthia Pollock, *Protecting Life on Earth.* Worldwatch Paper 87, December 1988.

Sloane, Christine S., and Tesche, Thomas W., eds., *Atmospheric Chemistry: Models and Predictions for Climate and Air Quality.* Chelsea, Mich.: Lewis Publishers, 1991.

INDEX